# WEATHER SENSITIVITY AND SERVICES IN SCOTLAND

# WEATHER SENSITIVITY AND SERVICES IN SCOTLAND

Proceedings of a Symposium held at the
University of Stirling, 4–5 February 1988

*edited by*

S. J. HARRISON and K. SMITH

Department of Environmental Science
University of Stirling

Scottish Academic Press
Edinburgh

Published by
Scottish Academic Press Ltd
33 Montgomery Street, Edinburgh EH7 5JX

First published 1989
ISBN 7073 0555 1

© 1989 Scottish Academic Press

All rights reserved. No part of this publication may be reproduced, stored in a retrieval system, or transmitted, in any form or by any means, electronic, mechanical, photocopying, recording or otherwise, without the prior permission of Scottish Academic Press Ltd

**British Library Cataloguing in Publication Data**

Weather sensitivity and services in Scotland
  1. Scotland. Weather
  I. Harrison, S. J. (Steven John), *1947–*
  II. Smith, K. (Keith)
  551.69411

ISBN 0-7073-0555-1

Printed in Northern Ireland at The Universities Press, Belfast.

# List of Contents

*Introduction*

1. Weather sensitivity in Scotland . . . . . . . . . . 3
   K. Smith, Professor of Environmental Science, University of Stirling.

*Section One*: ADVANCES IN OBSERVING AND FORECASTING THE WEATHER

2. Use of weather radar and satellite data in weather forecasting . . . 23
   C. G. Collier, Meteorological Office, Bracknell.

3. The use of numerical models in weather forecasting: achievements and prospects . . . . . . . . . . . . . . . 40
   B. Golding, Meteorological Office, Bracknell.

4. *In-situ* meteorological monitoring: the practical problem . . . . 55
   R. E. W. Pettifer, Managing Director, Vaisala (UK) Ltd.

*Section Two*: TRANSPORT

5. Meteorological Office services to transport . . . . . . . 65
   J. G. Allardice, Meteorological Office, Glasgow.

6. The response of a Regional Roads Department to adverse weather . . 75
   D. L. Brinham, Assistant Director, Roads and Transportation, Central Regional Council.

7. The effect of severe weather on the Scottish rail system . . . . 83
   C. L. Crawford, Chief Controller, ScotRail, Glasgow.

*Section Three*: AGRICULTURE, WATER AND WIND RESOURCES

8. Services to agriculture: past, present and potential . . . . . 93
   B. A. Callander, Meteorological Office, Bracknell.

9. Water resource management and flooding . . . . . . . 102
   R. J. Sargent, Chief Hydrologist, Forth River Purification Board, Edinburgh.

10. The meteorological needs of the wind turbine industry . . . . . 110
    G. Elliot and S. M. Barton, National Wind Turbine Centre, National Engineering Laboratory, East Kilbride.

*Section Four*: INDUSTRIAL APPLICATIONS

11. Sources of United Kingdom weather data . . . . . . . 121
    F. Singleton, Meteorological Office, Bracknell.

*List of Contents*

12. The application of Scottish weather records . . . . . . . 128
    R. C. Tabony and P. A. D. Brown, Meteorological Office, Edinburgh.

13. The effects of adverse weather on the construction industry . . . . 134
    J. R. T. Carson, James Miller Construction.

14. Forecast services for the building and construction industries in Scotland . 139
    H. Cumming, Meteorological Office, Aberdeen.

15. Weather interference with construction operations: Met Office climatological services . . . . . . . . . . . . . 143
    M. J. Prior, Meteorological Office, Bracknell.

16. Weather sensitivity in the gas industry . . . . . . . . 149
    R. A. Steel, Grid Controller, British Gas, Scotland.

17. Tourism and the Scottish weather . . . . . . . . . 162
    B. Hay, Research Manager, Scottish Tourist Board.

*Conspectus*

18. Weather sensitivity and the future for meteorological services . . . 169
    S. J. Harrison, Lecturer in Environmental Science, University of Stirling.

# Preface

THERE is growing evidence that Scotland, like many other countries, is more vulnerable to its own weather and climate than is commonly accepted. Even the routine variability of atmospheric conditions imposes large economic and social costs. More extreme events can seriously disrupt everyday life and often produce public criticism of the effectiveness of plans for coping with adverse weather. As an awareness of these burdens associated with atmospheric variability increases, so does the incentive to reduce the losses through the better application of weather and climate information. In turn, this trend is producing an upsurge in demand for improved atmospheric data and meteorological services.

The Climatic Hazards Unit at the University of Stirling pursues both pure and applied research into the impact of weather and climate on human society. The Meteorological Office is the government agency charged with the provision of public weather services in the United Kingdom. It was perhaps inevitable, therefore, that these two bodies should jointly organise a symposium designed to explore weather sensitivity and to demonstrate the benefits of information services, both existing and planned. Because weather and climate tend to assume regional characteristics and produce social and economic impacts which can be most easily interpreted on a national basis, it was concluded that Scotland provided a convenient scale at which to illustrate many of the relationships involved.

The symposium meeting, attended by almost 150 delegates, took place at the University of Stirling in February, 1988. The purpose of the symposium was not only to demonstrate the important role which weather and climate information already plays in the successful management of selected weather-sensitive enterprises but also to encourage greater dialogue between the suppliers and consumers of weather services in order to mitigate weather sensitivity in the future. It is felt that the supply of more relevant and timely information is already the major task facing the meteorological community. The effective use of such information by planners and policy-makers is likely to emerge as one of the social challenges of the next decade. By linking contributions from meteorologists and managers, it is hoped that this book will make a modest contribution to these important and practical goals in Scotland and elsewhere.

This volume is organised into six parts. Following an introductory overview of weather sensitivity in Scotland, *Section One* deals with recent advances in atmospheric science related to observing and forecasting the weather. Emphasis is on high technology remote sensing techniques, including the weather radar shortly to come to Scotland, and numerical modelling of the atmosphere for more specific forecasting purposes, although a cautionary note is also added about the use of more traditional weather observations. *Section Two* is concerned with the weather sensitivity of transport and shows how new technology and better applications of forecast information, which have been largely pioneered in Scotland, are already helping road and rail authorities to combat adverse weather conditions, especially

*Preface*

during the winter months. In *Section Three* the spotlight turns on the natural environment with an assessment of the atmospheric background to agriculture and water resources in Scotland, and provides an estimate of the reduced flood losses which might be expected with the advent of weather radar. It is shown that wind energy is a resource as well as a hazard in Scotland and a plea is made for more accurate observation of this asset. *Section Four* addresses the application of archived and forecast weather information for selected industrial purposes, notably for building and construction, gas supply and tourism. This section highlights the wide range of success currently achieved, with a detailed account of weather-related operations in the gas industry contrasting with the more general understanding of weather effects on tourism and recreation. Finally, an attempt is made to review the field and point the way to the future.

It would have been impossible to hold the symposium and to produce this book without the cooperation and support of many organisations and individuals. The organising committee, based in Stirling, was heavily dependent on advice from the Met Office and, in particular, the editors wish to acknowledge the help provided by Dr P. Ryder, Mr J. G. Allardice and Miss M. Roy. Dr Ryder, Deputy Director of the Met Office, also chaired one of the symposium sessions, as did Mr R. McGillivray, Chief Engineer, Scottish Development Department. The willingness of all the contributors, and their parent bodies, to participate has been much appreciated. Thanks are also due to various staff at the University of Stirling for their assistance at all stages of the symposium and subsequent publication of the book. The editors are especially grateful to staff in the Department of Environmental Science; Mrs I. Mack and Mrs E. Urquhart (secretarial services) and Mrs M. Smith and Mr K. Dockery (drawing office services).

S. J. HARRISON.  
K. SMITH

July 1988

University of Stirling

# INTRODUCTION

# Weather sensitivity in Scotland

## K. SMITH

*Department of Environmental Science, University of Stirling*

### INTRODUCTION

FLUCTUATIONS of weather and climate have always been important to Scotland. In the past, extreme atmospheric conditions have led to notable weather-related disasters such as the collapse of the Tay Bridge in 1879 and the 'Glasgow hurricane' of 1968, which killed nine people and badly damaged some 70,000 local authority houses. Today, even in 'normal' conditions, there are few aspects of Scottish life that are not weather sensitive.

Many of Scotland's major industries and economic activities, such as agriculture, transport or tourism are highly sensitive to atmospheric variations on all time-scales. A significant percentage of Scotland's economy is directly or indirectly used to mitigate the socio-economic effects of such variability. Many industries such as energy and water supply or insurance exist, at least in part, to smooth out the irregular burdens which the atmosphere imposes on society. Some activities are well adapted to atmospheric variability but, often, poor planning and social attitudes have tended to obscure some of the relationships and preclude the effective use of weather and climate information to reduce weather sensitivity.

Such a situation is far from unique. Indeed, what has been termed one of the more striking science policy developments of the past decade has been the formulation and first-phase implementation of 'National Climate Programmes' in many western countries (Lamb *et al.*, 1985). These programmes recognise that atmospheric fluctuations can have major influences on food, water, energy supplies and other conditions relating to national security and human welfare. They typically seek to combine existing basic atmospheric research and monitoring programmes with efforts to assess the impact of climate. Consequently, most of the climate programmes, as in the USA, specifically aim to identify '... procedures to evaluate climate's effects on society, the economy and the environment in order to develop responses and strategies for dealing with climatic fluctuations' (National Oceanic and Atmospheric Administration, 1980). There is general agreement that the second of these goals (the development of 'responses and strategies for dealing with climatic fluctuations') will be more difficult to achieve than its necessary forerunner. At the present time, no such coordinated programme exists in the UK, let alone in Scotland.

Atmospheric variability is a difficult concept to define. As Parker and Folland (1987) have stated, there is no clear boundary between short-term variability consisting of spells, months or seasons of differing character, which may well be part of a fixed climate, and long-term variability, which involves changes in the frequencies and mean values of climatic quantities (temperature, rainfall) over periods of several decades or longer. Weather sensitivity deals essentially with the consequences of short-term variability and most of the relevant fluctuations operate well within the annual timescale. Indeed, the variability important for weather sensitivity often falls during the time-frame within which useful model predictions and weather forecasts are currently made (usually about a week). Longer-term variability concerns all other periods for which, at the present time, it is necessary to predict atmospheric behaviour on the basis of a more statistical or actuarial approach.

## WEATHER SENSITIVITY

Weather sensitivity in Scotland, as elsewhere, is determined by the balance between what are perceived as atmospheric 'resources' and atmospheric 'hazards'. These terms reflect personalised value judgements which are not always helpful because the atmospheric processes responsible for such events are neither benign nor hostile but entirely neutral.

Fig. 1.1 shows the arbitrary boundaries conventionally imposed on atmospheric variability to identify resources and hazards. Most social and economic activities are geared to some expectation of the 'average' conditions. As long as the variation of any element remains fairly close to this expected mean, the activities are well adjusted to the conditions, insignificant damage occurs and the atmosphere may be regarded as a resource. However, when variability exceeds some perceived

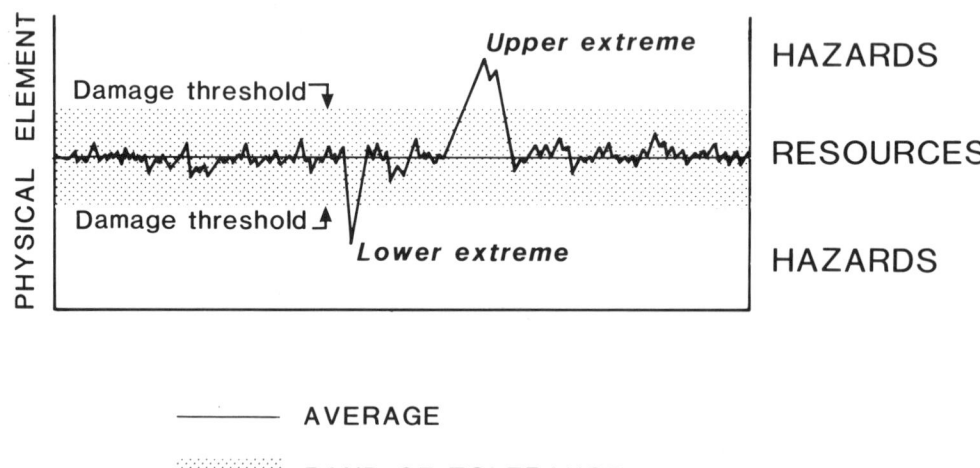

FIG. 1.1. Weather sensitivity expressed as a function of atmospheric variability and socio-economic tolerance. (Modified from Hewitt, K. and Burton, I. (1971) *The Hazardousness of a place,* University of Toronto Press.

threshold, that same element is immediately regarded as a hazard and, often wrongly, is termed an extreme event. In practice, the hazards merely represent the unexpected (or unprepared for) outer limits of a distribution that would otherwise be recognised as a resource. Frequently, there is only a very fine line distinguishing the two—between water under control in a reservoir and out of control in a flood or between snow as a resource on the Cairngorm ski slopes and as a disbenefit on the approach roads.

Although weather sensitivity may be seen as some function of physical exposure, measured by atmospheric variability, and social vulnerability, measured by the tolerance between upper and lower damage thresholds, comparatively little is known about the relationships. Often it is not known which weather element, or combination of elements, is of most importance for a particular enterprise. The threshold values are usually poorly defined and may well be better represented by a transition zone than a sharp boundary. Frequent and unpredictable low-level variability around a very critical threshold may well be more significant than isolated or more predictable extremes. For example, 0°C is a critical threshold because of the freezing of water at that temperature but, even in the Scottish winter, such a value could hardly be described as an extreme. Equally, the extremes recorded in Scotland are nothing more than the normals experienced elsewhere on the globe. 'Normal' variability may, therefore, exert a greater toll than the rarer, more extreme, events. Thus, the economic loss to Scottish forestry associated with endemic windthrow is substantially greater than that from catastrophic damage (Atterson, 1980).

The consequences of atmospheric variability upon environmental and human systems can be quite complex. Rarely does a straightforward cause and effect situation apply. At the very least, it is more usual to have an ordered cascade of impacts on a particular sector or activity which is likely to range right through both environmental and social systems, e.g. from biophysical effects through to economic impacts. For example, a dominantly cyclonic summer characterised by low temperatures, limited sunshine, high winds and heavy rainfall (perhaps similar to 1985) will not only have an ordered sequence of impacts on, say, agriculture ranging through waterlogged soils, reduced crop yields, additional costs and reduced farm incomes. It will also impact widely throughout the Scottish economy affecting both natural factors such as loch levels, flooding, fisheries and activities and costs as diverse as recreation, tourism, construction, insurance, road accidents and retailing.

The relationship between exposure and vulnerability is further complicated by the fact that both entities are dynamic and may change independently through time. Fig. 1.2 shows just three idealised time-series of atmospheric variability and the social band of tolerance which lead to an increase in hazards relative to resources. Pattern A represents a constant band of tolerance, with unchanging variability, but a downward trend of mean value. B also has a constant band of tolerance, together with a constant mean value, but increasing variability. Finally, pattern C shows constant physical exposure, represented by unchanging variability and mean, but a narrowing band of tolerance.

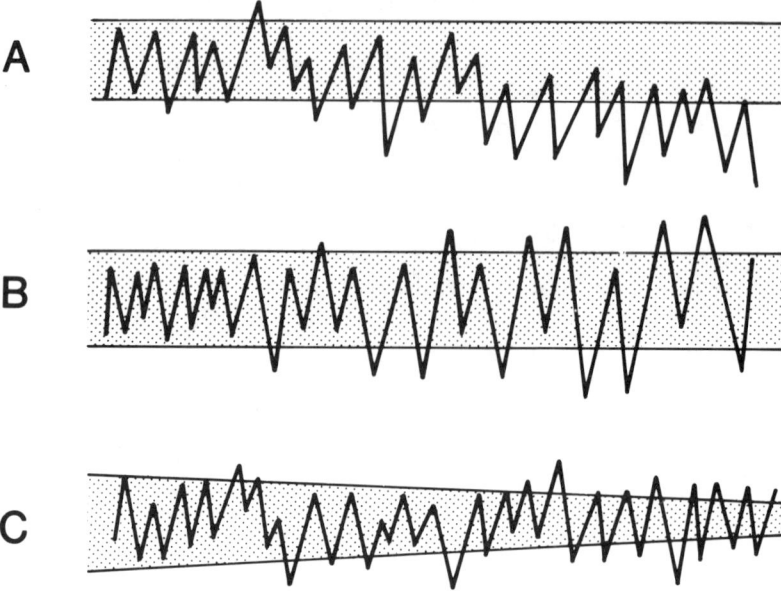

FIG. 1.2. Idealised time-series of atmospheric variability and band of tolerance showing changes which can lead to increased weather sensitivity.

## PHYSICAL EXPOSURE IN SCOTLAND

With respect to latitude and average altitude alone, it might well be expected that Scotland suffers a harsher climate than the rest of the UK and that certain economic activities are more marginal. For example, Scotland generally has a shorter growing season and a longer domestic space-heating season than is found elsewhere in the country. In combination, this often ensures more extreme conditions than elsewhere. Scotland holds the record for extreme gusts of wind in the UK at both a low-level site (118 knots at Kirkwall) and a high-level site (125 knots on Cairngorm). Exposure to strong winds is probably the most obvious climate stress arising directly from Scotland's geographical position.

Over the 50-year period 1937–1986, Scotland also experienced the lowest absolute minimum air temperature recorded at any site in the UK during no less than 43 of the years. Most of these low temperatures have been observed in highland straths (notably those of the upper Dee and the Spey) under inversion conditions in winter when the effects of large-scale shelter created by the amplitude of relief are more significant than mean altitude (Tabony, 1985). Smaller-scale relief effects regularly

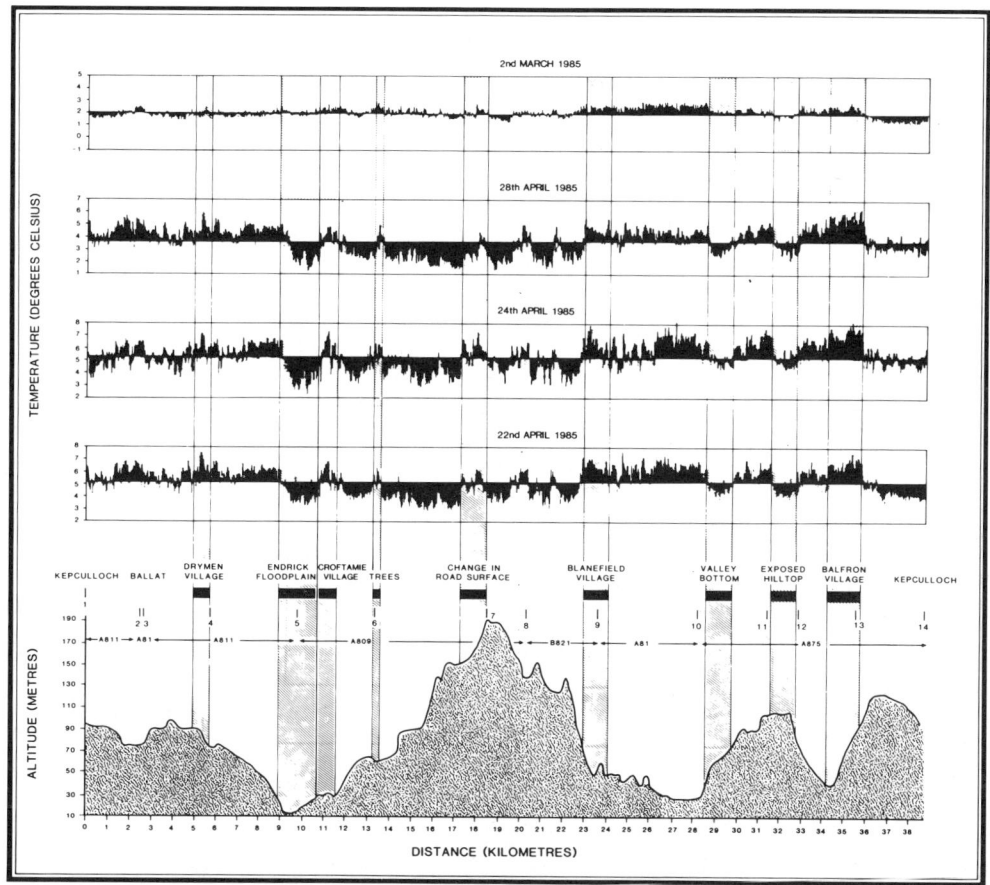

FIG. 1.3 Road surface temperatures recorded in Central Scotland on four nights during 1985 with a cross-section of the route. (From Smith, 1988).

occur everwhere to produce marked spatial variability of minimum temperatures. Fig. 1.3 shows characteristic nocturnal 'thermal fingerprints' for part of the rural road network in central Scotland. An intimate relationship between road surface temperature and altitude, shelter and exposure can be seen with sharp contrasts in temperature of up to 5°C occurring between low-lying frost pockets and adjacent small urban heat islands (Smith, 1988a).

However, the temporal variability of atmospheric conditions is a more important determinant of weather sensitivity. Fig. 1.4 depicts the long-term (55 year) mean cycle of daily temperatures through the year together with the pattern for 1987 at a representative station (Craibstone in north-east Scotland). Even the long-term average shows irregularity over periods of a few days. Marked fluctuations ranging up to a month's duration are evident in the 1987 trace where there is a clear contrast

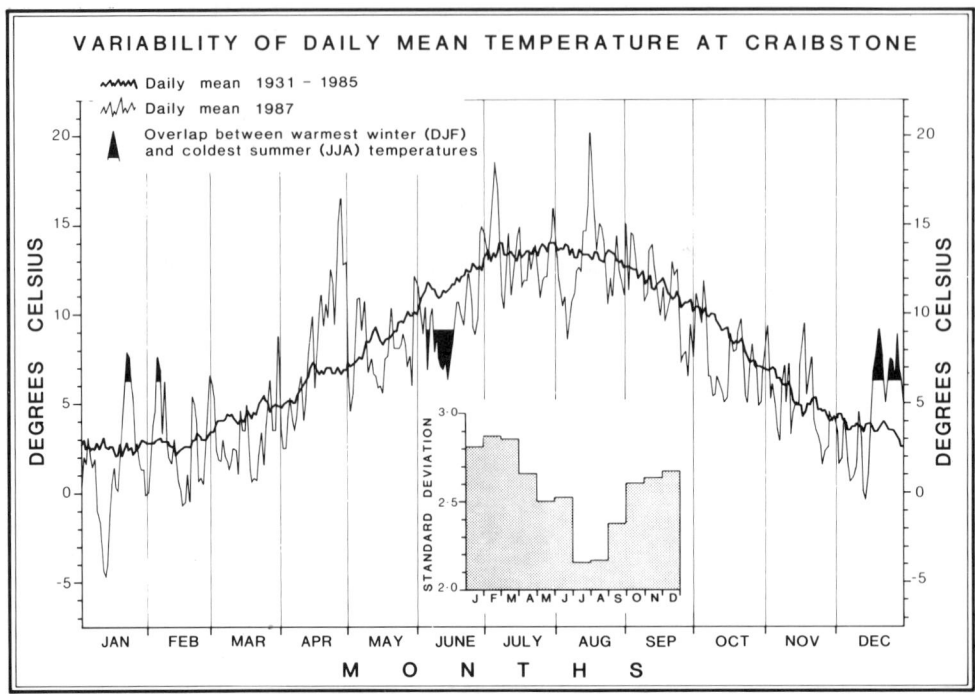

FIG. 1.4 Mean daily air temperatures at Craibstone, near Aberdeen, during 1987 with the 1931–1985 normals. The shaded areas shows the overlap between the warmest winter (DJF) and the coldest summer (JJA) temperatures. The inset standard deviation graph refers to the long-period data.

between early and late January together with that between the cold March (1·2°C below the 1951–80 average) and the warm April (2·0°C above the equivalent norm). December is also notable for strongly contrasting spells. As shown, the mean temperature can vary by 5–6°C on adjacent days, especially in winter. It is also a curiosity of the Scottish climate that, in individual years, the warmest day of winter (DJF) is warmer than the coldest day of summer (JJA) and the overlap in 1987 is indicated.

Winter variability of temperature bears most heavily on weather sensitivity for at least two reasons. Firstly, as can be seen in Fig. 1.4, variability is greatest at this season. At Craibstone the 55-year standard deviation for daily means ranges from 2·874 in February to 2·162 in July. Secondly, winter temperatures are often marginal with regard to the freezing point. Fig. 1.5 shows mean minimum air temperatures for 1940–70 over north-western Europe in January. Although most of continental Europe is colder than even the highest parts of Scotland, the corollary is that all of Scotland—and especially the settled lowlands—its clearly more marginal and less predictable with respect to the zero degrees threshold. This has significance for a number of economic activities, not least road transport. As Thornes (1987) has

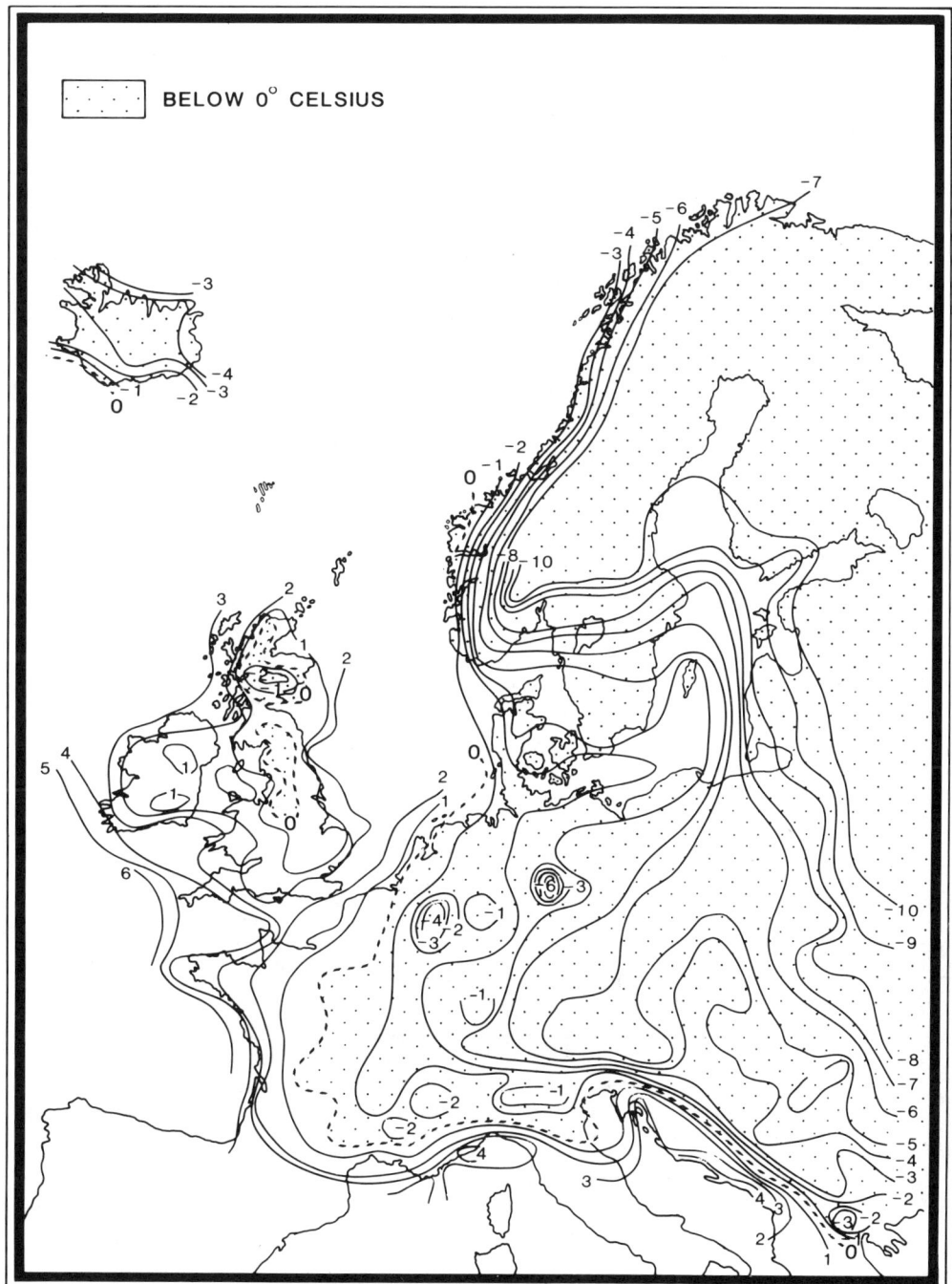

FIG. 1.5. Mean minimum January air temperatures over north-west Europe during the 1940–70 period. (From Thornes, 1985).

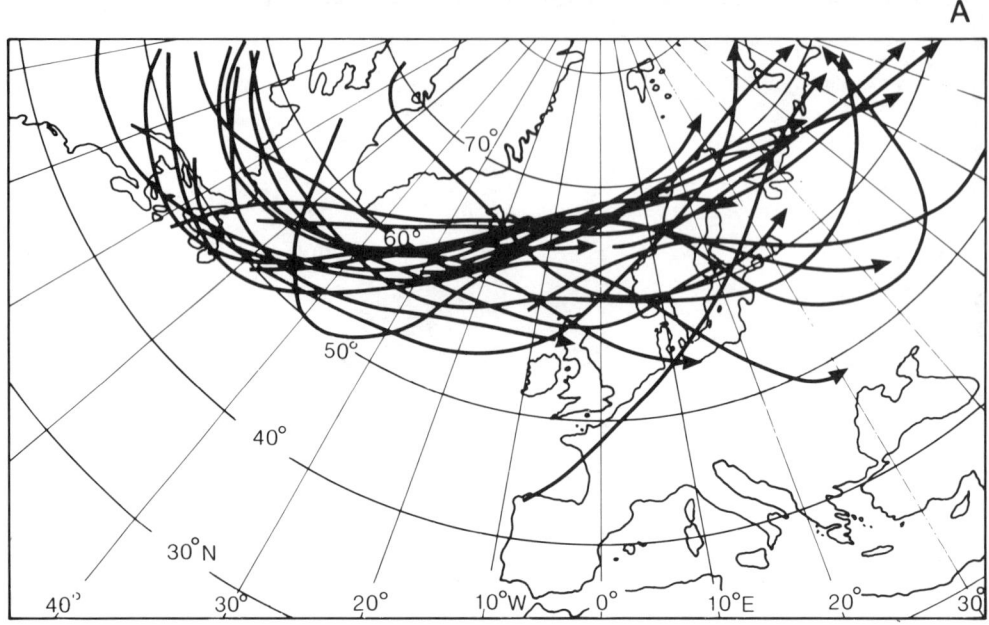

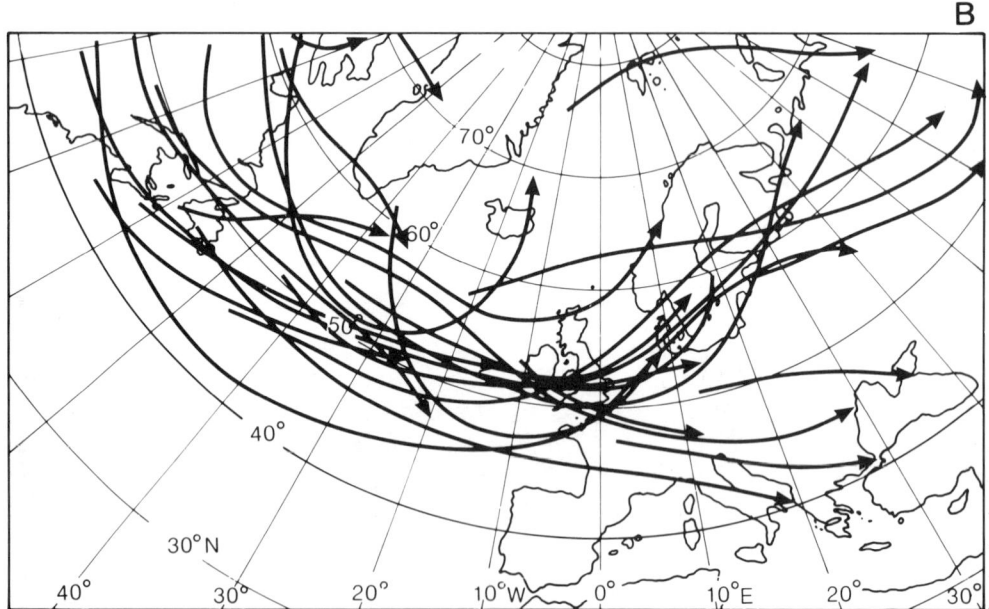

FIG. 1.6. Position of 5-day (pentad) mean maximum jet stream speeds (>50 knots) at 500 mb during June, July and August for (A) 1983 and (B) 1985. (From Folland and Woodcock, 1986).

pointed out, ice is most slippery around 0°C when skid resistance may be lowered by a thin layer of lubricating water. On average, icy roads are thus more dangerous in Scotland than in colder climates where it may be possible to drive safely on a snow-covered surface at much lower temperatures.

Similar winter problems of marginality apply to snow, which usually falls with ambient temperatures close to the freezing point. As Hunt (1987) has pointed out, the situations which lead to widespread snowfall usually involve the fusing together of cold and mild air, thus causing uncertainty as to whether snow or rain should be forecast. What snow does fall is often wet and sticky. Even when snowfall can be confidently predicted, there is the further problem of deciding whether it will lie at all levels or just remain on higher ground. A drop in temperature of only one 1°C in winter can make all the difference between a wet day and a day of travel chaos.

Parker and Folland (1987) have drawn attention to the importance of longer spells of more or less persistently abnormal weather which may be superimposed on the short-term variability. This is because the tendency for a given type of long spell to repeat at a given season, or even over more extended periods, more frequently than before would constitute one prominent type of regional climatic change. Such fluctuations reflect changes in the general atmospheric circulation such as the location of the jet stream which, in turn, influences the path of mid-latitude depression systems. Fig. 1.6 contrasts the jet-stream tracks at about 6000 metres above sea level for 5-day periods during the two very different summers of 1983 and 1985 (Folland and Woodcock, 1986). During 1983 these tracks steered most Atlantic depressions well to the north of Britain. Consequently, Scotland enjoyed a predominantly anticyclonic summer with warm, dry conditions. In 1985 the tracks were mainly over England and, because the wettest conditions are generally slightly to the north of the main jet axis, Scotland experienced a very rainy summer. At Stirling (Parkhead), for example, summer (JJA) rainfall in 1983 was 52 per cent of normal compared with 218 per cent of the average in 1985. The effects of the 1985 summer on Scottish agriculture alone were little short of disastrous with direct weather-related losses estimated between £150 and £200 million (Parker *et al.*, 1986).

The water supply industry is well-adjusted to short-term weather extremes but may be second only to agriculture in its sensitivity to the longer-term atmospheric variations. Even in such a comparatively well-watered country as Scotland, shortages of rainfall can create problems. The relatively high dependence on surface water supplies, fed by rivers which are very small in comparison with those in many other countries, means that Scotland's water resources rapidly show the effects of any rainfall deficiencies. This is especially true if the short-fall in precipitation accentuates the regional differences between west and east (Smith, 1977). For example, during the early 1970s a succession of dry years produced totals in eastern Scotland well below the normal expectation. Fig. 1.7 shows the situation in 1972 when the national average rainfall was 82 per cent of the long-term mean but parts of the Moray Firth and the Lothians had only around 60 per cent of the average (Scottish Development Department, 1973). By March 1973 river flows in the

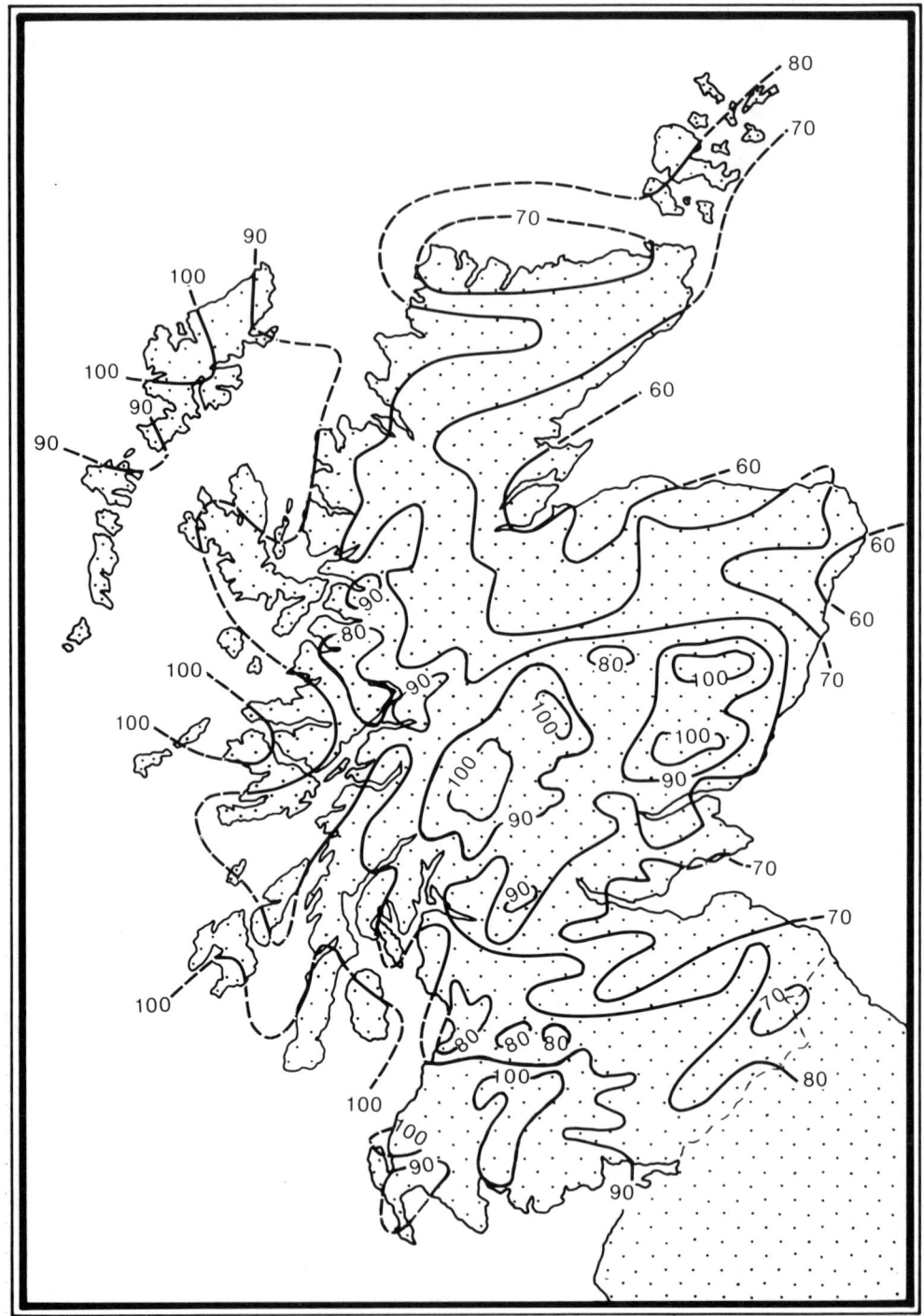

FIG. 1.7. Distribution of rainfall over Scotland in 1972 as a percentage of the 1916–50 normals. (From Scottish Development Department, 1973).

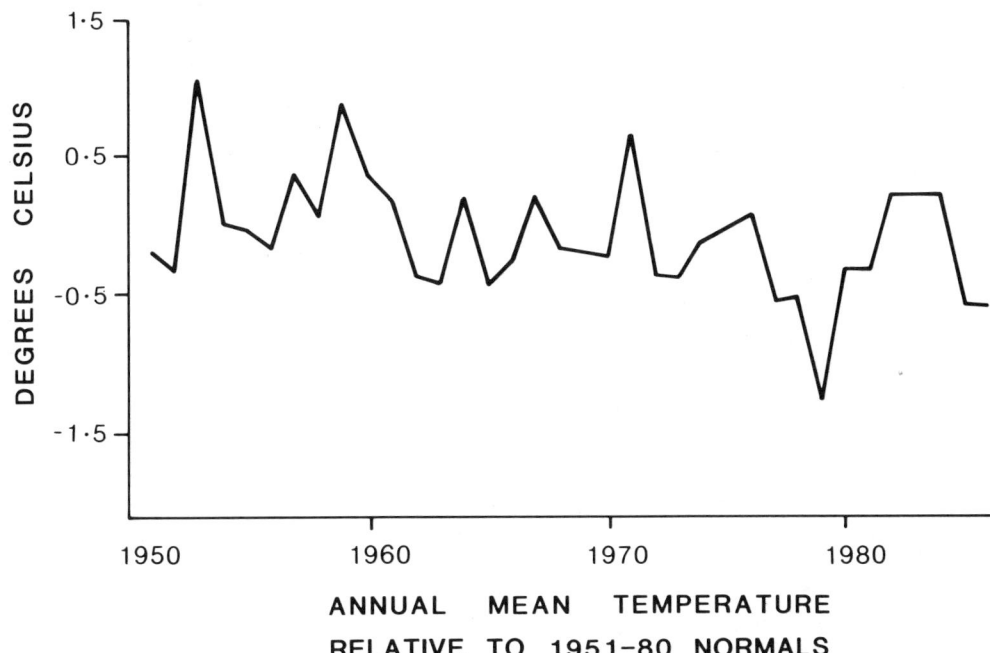

FIG. 1.8. Mean annual air temperatures for northern Scotland from 1950 relative to the 1951–80 normal showing a declining trend since 1960. (From Parker and Folland, 1987).

Lothians and South East had fallen to between 12 and 20 per cent of normal whilst the reservoirs supplying St Andrews and Burntisland were at only 25 and 19 per cent of full capacity respectively.

The above example illustrates the socio-economic importance of long periods of generally abnormal weather which may extend well beyond individual seasons or years. This is especially so if a trend emerges which threatens to breach the tolerance thresholds. For example, a definite cooling trend has been observed in north-western Britain since about 1960 which appears especially marked in northern Scotland (Parker and Folland, 1987). Fig. 1.8 summarises data from thirteen stations north of latitude 57°N and indicates an overall reduction of around 0·5°C during approximately the last 25 years which may be related to the simultaneous decline in sea surface temperatures in the mid-latitude north Atlantic over which airstreams often have to pass to reach this area. The implications of this cooling for activities such as agriculture have yet to be assessed.

## SOCIAL VULNERABILITY IN SCOTLAND

This concept is less well understood than that of physical exposure. It reflects those factors in the economic and social fabric which determine exposure to atmospheric variability and, in particular, any ongoing changes within a community or area that make it more vulnerable to weather and climate impact. Scotland shares many broad characteristics with the rest of the UK and it is difficult to identify specific regional factors which lead to a narrowed band of tolerance.

One characteristic, however, is that almost 85 per cent of the population is concentrated in the central belt. Despite the development of new towns, much of this population suffers from urban deprivation and is housed at comparatively high densities in local authority property which appears especially vulnerable to the elements. For example, in the 1981–82 winter some 23,000 Glasgow District properties (approximately one in eight) were damaged by frost at an estimated cost of £13 million. On the other hand, Scotland also has a large number of isolated rural communities which are in danger of being cut off by severe weather in winter. Remote areas rarely have rail links as an alternative to the road network and the disruption arising from heavy snowfalls has been well documented (Perry and Symons, 1980; Perry, Symons and Williams, 1984).

Beyond this, Scotland exhibits certain general trends which tend to increase weather sensitivity. Some of these relate to rising social expectations. For example, people have become more mobile in recent years and expect to get themselves, or their goods, transported around the country quickly and safely no matter what the weather conditions may be. According to McAlonan (1984), the suburban commuter now expects a level of local authority service which produces black road surfaces each morning irrespective of the weather. The same absolute security of service also appears to be expected by consumers from other weather-dependent enterprises such as energy supply and water supply. This spiral of increasing expectation of a fail-safe performance can lead to serious disruption when failure occurs.

The Scottish economy functions on the basis of considerable interdependence between the individual areas and sectors so that, for example, a failure of the transport system is immediately felt in the distributive and retail trades. The electricity industry fulfils a key role in that any failure of supply, perhaps created by storm damage to overhead transmission links, would cascade throughout the system from primary production to the service and financial sectors. The implications of increasingly complex infrastructures are compounded by the effects of the more recent trend to greater efficiency in commerce and industry. Frequently the drive for increased competition has resulted in reduced manning and smaller operating margins. In turn, these allow less scope for responding to atmospheric variability.

Growing prosperity in recent decades has greatly increased the exposed risk for weather-related property damage. Many households now have central heating systems with the greater risk of damage from burst pipes during spells of winter frost. The upward trend in car ownership, together with the increasing sophistication of the vehicles, has contributed to a steep rise in the cost of weather-related road accidents.

One of the most basic social failures in reducing weather sensitivity has been an inadequate appreciation of the challenges and opportunities involved. This may have resulted from some combination of assumptions by the managers of weather-sensitive enterprises, such as that atmospheric conditions will be average, or that nothing can be done about the weather, or that information is unavailable, or inaccurate or expensive. Sometimes the failures are institutional rather than individual. For example, the Met Office has only recently started to market its services in a fully professional manner. Many business firms appear to be dominated by short-term goals and what are often conveniently termed 'avoidable costs'. This often means neglect of future weather planning in favour of a more immediate maximisation of investment returns.

## EMPIRICAL EVIDENCE OF WEATHER SENSITIVITY

Comparatively few regional assessments of weather sensitivity are available, partly due to the lack of convenient data sources. Content analysis of newspapers represents a possible source of comprehensive, long-run, information and over 3000 issues of the daily morning *Glasgow Herald* were examined over a recent 10-year period to determine the incidence of weather hazards within a 40 km radius of the city of Glasgow (Smith, 1985). It was found that 6 per cent of all newspaper issues contained specific reports of adverse weather impacts, with about 80 per cent occurring in the six winter months (October–March inclusive). The three most frequent hazards were frost, gales and snowstorms. The most important conclusion was the vulnerability of transport, especially road transport, to severe weather disruption. This confirms previous work in North America (Rooney, 1967; Defreitas, 1975; Bertness, 1980). One implication is for weather-related road accidents, and wet weather alone is sufficient to increase personal-injury road accidents by some 23 per cent in Glasgow (Smith, 1982a).

Another approach to weather sensitivity is through the demand for weather information. This approach assumes that a more practical test of atmospheric impact depends not so much on how sensitive particular operations may be to the weather, but rather on what managerial action may be possible either to minimise weather-related losses or increase weather-related gains. In some cases, no amount of weather information will help but, in many instances, weather information can be employed to advantage. As indicated by Houghton (1987), sensitivity to weather information depends not only on weather sensitivity but also on the availability of appropriate advice and on the ability of the consumer to use the advice.

In a short series of papers, Smith (1981, 1982b, 1983) examined the weather sensitivity of spontaneous telephone enquiries from the general public to the Glasgow Weather Centre. These calls, all seeking current or forecast weather advice, were classified into twelve categories and could be analysed on various timescales. Five of the larger categories, accounting for over 90 per cent of the total enquiries, showed some weather sensitivity. These categories were agriculture, road transport, marine, holiday and building enquiries. Subsequent work concentrated on road transport and building enquiries to determine if specific meteorological thresholds led to a significant increase in the demand for weather information. In the case of road transport it was found that adverse driving conditions in winter, represented by differing measures of snowfall, fog and heavy rainfall, produced at least a two-fold increase in enquiries over timescales from one day to one hour. Snowfall had the greatest effect and was capable of increasing demand around six-fold. In the case of building enquiries, peak demands for weather information were similarly linked to adverse meteorological thresholds and this analysis provided indirect confirmation of the 'prohibitive values' suggested by Prior and King (1981) as being likely to stop outdoor work on building sites. More generally, these studies showed not only that there are large sectoral differences in the sensitivity of demand for weather information but also that the variations in demand can be related to the severity of the prevailing weather conditions.

## THE ECONOMIC DIMENSION

It is not sufficient to demonstrate weather-sensitivity for different sectors, or even to quantify weather-related costs. This is because the most important practical issue is not what the weather *costs* but rather what the consumer can *save* as a result of taking weather information. Weather services are of value only if they are incorporated into the decision-making process and reduce weather-related costs. Thus, a proper understanding of weather sensitivity must include some knowledge of how the user actually incorporates weather information into the decision-making process and an awareness of the cost:benefit implications. Some activities, like the gas supply industry, are so weather dependent on a day-to-day basis that they have well-rehearsed routines for using the information (Steel, 1988, this volume). Other industries also have particular problems which, on cost:benefit grounds, justify them taking a contract service from the Met Office. For example, ScotRail has recently elected to purchase a service tailored more closely to its requirements at an annual cost of £10,000. This should be seen in the context of the estimated £3·5 million running cost of an average winter for British Rail and isolated effects such as the cold weather of January 1987 which created revenue losses of about £15 million throughout the national network (Worrall and Pitts, 1987).

Perhaps the sector which has gained most in recent years from the rational application of improved weather information in Scotland has been winter road maintenance. The average annual bill for snow and ice control on Britain's roads is now about £120 million and it has suggested that the burden of costs is especially heavy in Scotland (Edmond, 1985). In Highland Region particularly it appears that, at least from the mid-1970s, expenditure on winter maintenance increased rapidly and, as shown in Fig. 1.9, until 1983–84 took a growing proportion of the total road maintenance budget. By the following year, Highland had become the first Regional Council in Scotland to introduce the new technology available for ice detection and the forecasting of minimum road surface temperatures (Thornes, 1985). Initial savings of at least 20 per cent in road salting have been claimed for the system and it is tempting to interpret the subsequent downturn in relative winter expenditure as a direct reflection of these developments.

Unfortunately, there are many other weather-sensitive enterprises where the

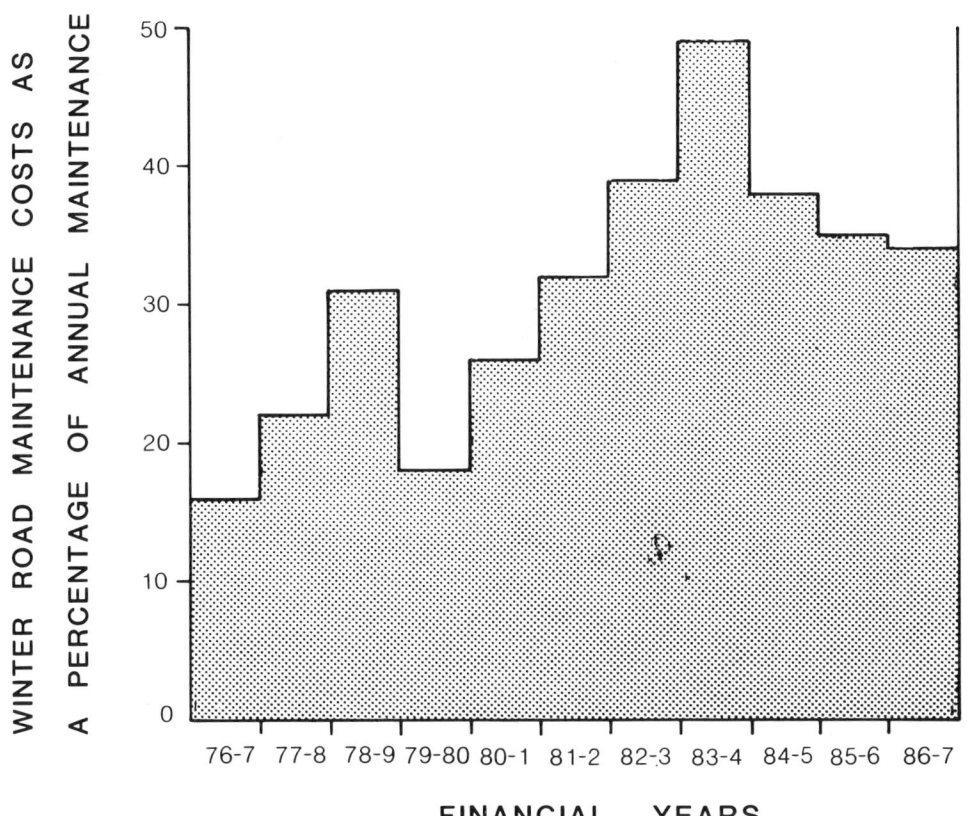

Fig. 1.9. Annual expenditure on winter road maintenance by Highland Regional Council as a percentage of total road maintenance costs.

value of, and potential applications for, weather information is less well understood. A major problem is that the relevant managers and decision-makers are often unaware of the avoidable weather costs or of the specialised information which is now becoming available. There is a special need to know how the relative scale of the meteorological factors to be included in the process compares with the other uncertainties inherent in the analysis. This implies the parallel development of weather forecasting and economic forecasting. Maunder (1986) has recently demonstrated many of the complexities of making weather-sensitive decisions under conditions of uncertainty and has also shown how difficult it is for weather and climate information to establish its role in the market place. In some instances, commercial confidentiality may be the prime reason for concealing cost:benefit estimates.

Beesley and Budd (1987) have outlined a possible approach to optimal weather protection (Fig. 1.10). The vertical axis measures costs and benefits whilst the horizontal axis represents increasing severity of weather conditions. It is assumed

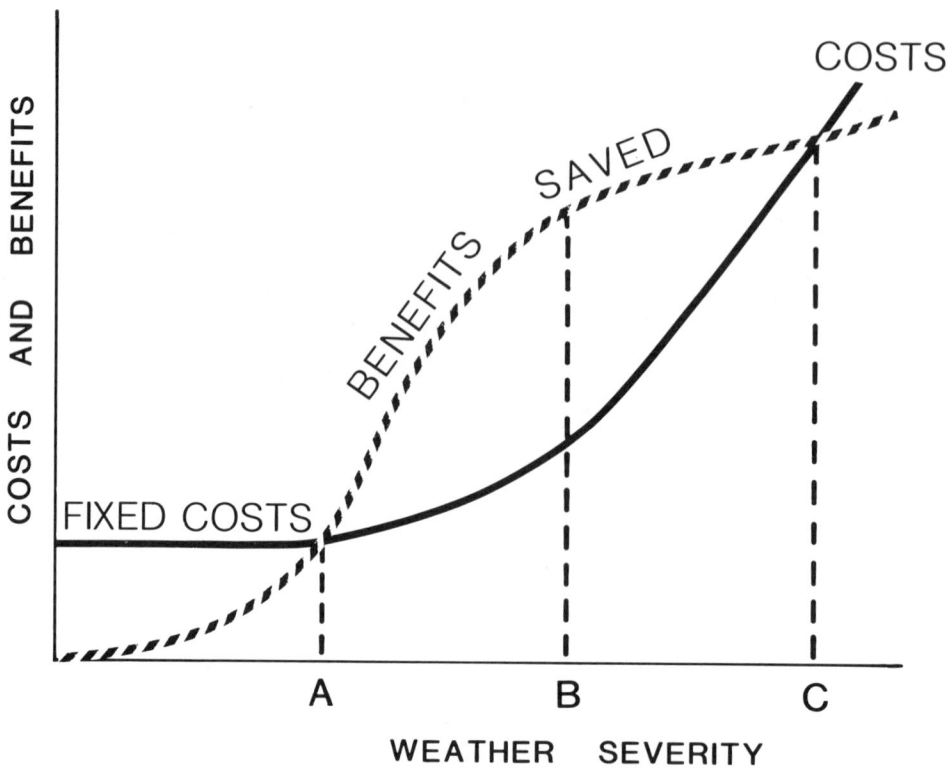

A = expenditure not justified

B = optimal expenditure (marginal benefit = marginal cost)

C = "break-even" point

Fig. 1.10. A possible allocation of costs and benefits involved in weather protection, with particular reference to transport. (From Beesley and Budd, 1987).

that there will always be some fixed costs to providing weather protection and that, beyond some meteorological threshold, the cost of maintaining normal working increases more rapidly. It is also assumed that the benefits saved increase rapidly beyond a certain point. Below the meteorological threshold A, there is no weather disruption and the cost of weather protection exceeds the benefits saved. Optimal expenditure occurs at point B where the marginal benefit saved is equal to the marginal cost of saving it. Point C is the break-even point beyond which total expenditure again begins to exceed the benefits saved. Hopefully, most operations already function at some point between A and C but the practical question is whether the cost:benefit ratio can be improved by either increasing or decreasing expenditure.

## THE WAY AHEAD

Progress towards a better adjustment to weather sensitivity in Scotland depends on a number of developments. Firstly, there is a need for improved weather information. This is dependent on better observations to support synoptic forecasting, such as are promised by the arrival of weather radar in Scotland, and also on the erection of specialised, sub-synoptic, data collection networks designed for particular needs. New technology now permits much cheaper real-time access to weather information, even from remote and inaccessible sites, and to climatological data bases for all who need it. More dedicated weather monitoring systems, similar to the existing network of automatic sensors currently deployed on Scotland's roads, are likely to be installed in future. More consumer-oriented weather forecasts are required together with greater use of the climatological archives to provide scenarios beyond present-day model predictions.

Secondly, new systems for delivering the information to the users need to be devised. Short-term information is currently available from a variety of sources depending on the conditions (local radio, British Telecom's Weatherline, the police etc). One possible approach might be a national viewdata system. This could be interrogated for specific forecast information by a wide spectrum of users and could also contain near real-time and immediate past data relevant to decision-making in business. Alternatively, if the commercialisation of European weather services envisaged by Tennekes (1988) actually occurs, it may be that the private sector will start to invest in the computing and interactive telecommunication systems required to transmit dedicated, value-added, products to individual clients.

Thirdly, there is a need to educate the appropriate managers and decision-makers to adopt a more positive attitude towards weather sensitivity. As Lamb (1981) has pointed out, it is necessary not only to identify the activities which are most affected by atmospheric variations, but also to determine which of the sectors or regions has the greatest managerial flexibility to take advantage of better weather information. At present, there is a lack of communication between pure/applied research expertise in meteorology and the economic sectors which are weather-sensitive. One way of overcoming this would be to convene user workshops for different key sectors such as agriculture, construction, tourism, (Harrison, 1988, this volume).

The potential for action is considerable. It may well be that such an ambitious agenda requires some coordination. Certainly, the Canadians, among others, believe that the cost of a comprehensive national climate programme is modest compared with the opportunities which it offers (Ferguson and Phillips, 1986). At the very least, the scale and complexity of the task suggests the need for greater cooperation between all parties (Met Office, universities, public agencies, private sector) interested in reducing weather sensitivity in Scotland. This would be especially opportune in view of the growing recognition of the need for integrated regional assessments of climate impact (Anonymous, 1987; Chen and Parry, 1987). A major reason is that, as in Scotland, weather and climate tend to assume regional dimensions. Moreover, within Scotland there are close links between the environmental and socio-economic systems which are often dissimilar from those elsewhere in the United Kingdom. The distinctive conditions in Scotland offer opportunities to develop a manageable and innovative research and development programme designed to promote more rational policy decisions to minimise weather sensitivity in the future. It will be interesting to see if these opportunities are grasped.

ACKNOWLEDGEMENTS. The author wishes to thank the Meteorological Office for the supply of data for Fig. 5 and for permission to reproduce Fig. 9. Central Regional Council kindly provided data for Fig. 4 and Highland Regional Council is also thanked for the information shown in Fig. 10.

# SECTION ONE

# ADVANCES IN OBSERVING AND FORECASTING THE WEATHER

# Use of weather radar and satellite data in weather forecasting

## C. G. COLLIER

*Meteorological Office, Bracknell*

## INTRODUCTION

WEATHER radar data have now been available for some forty years and data from satellites for over twenty-five years. However, it is only recently that these data have begun to have a significant impact on operational weather forecasting in situations other than for severe local storms and hurricanes. In the USA analogue data from radars and satellites have provided forecasters with extensive views of weather as it happens since they first became available, but systems which enable digital data from different sources to be blended and presented in real-time have been developed only comparatively recently.

Considerable work in many countries is now underway aimed at processing radar and satellite data to provide products which will aid forecasting at all lead times. Satellite instruments are capable of generating information which can give measurements of many atmospheric parameters. This paper concentrates on work in the UK Met Office which is used illustratively to demonstrate particular applications. Although mention will be made of future work, emphasis is on details of applications which are now contributing operationally. Hence, the extraction of winds from the ESA ERS-1 satellite scatterometer due for launch in 1990 is excluded. This paper demonstrates that we are now witnessing major advances in the data bases available for weather forecasting, which are expected to lead to improvements in forecast accuracy and utility.

## USES OF IMAGERY IN CONVENTIONAL FORECASTING

### Monitoring the atmosphere

Observations of precipitation and cloud cover allow more precise information on weather to be given, especially in otherwise data sparse areas. New short-period forecasting services are commerically important, but the needs of these services will be discussed separately as they are more specific. In general, forecasting involves three main stages:

(a) The state of the atmosphere is analysed using a combination of observations of different types including *in-situ* measurements (temperature and humidity at different levels in the atmosphere, and quantities measured regularly at the surface, such as pressure and type of weather), and remotely sensed observations (radar maps of precipitation, temperature profiles and images of clouds from satellites).

(b) The changes in the state of the atmosphere over a period of up to a few days

are predicted using numerical solutions of the mathematical equations that describe its evolution.
(c) The forecaster uses information from all the different sources to predict the likely weather resulting from changes in the state of the atmosphere. The details of the process will depend upon customer requirements. Thus, the forecaster will use the output from numerical forecasts, his experience of the evolution of similar systems in the past and the recent history of the system of particular importance to the current situation.

General forecasting covers both short-range (12–24 hours) and medium-range (a few days) services, but is regarded here as that element of forecasting which, whilst using numerically derived products, is essentially a manual operation involving expert human forecasters. Short-period forecasts (known as TAFs and TRENDs) and warnings are vital elements in forecasting for aviation, particularly in relation to aircraft safety. This applies both to the major airports with forecasting staff present and to other airports or helicopter landing sites serviced by forecasters located elsewhere. Forecasts of the timing, duration, and intensity of weather elements, particularly visibility and cloud base, are the essence of this type of forecasting. In addition, many ground-based and airborne activities can be impaired by lightning.

Forecasting for the general public via the media, public utilities (gas, electricity, water), transport (shipping, rail, road), and industrial concerns (agriculture and horticulture, civil engineering, oil and gas extraction offshore, and industrial processes including consumer demand) and pollution monitoring (acid rain, chemical and radioactive releases) is organised in a variety of different ways in different meteorological services. Often there is a division of responsibilities between a central forecast office and numerous small offices responsible for particular geographic localities or groups of users. However, the common need is for continuous monitoring of the state of the atmosphere on different spatial scales through the analysis of observations. The quality of such observations must also be continually checked in order that appropriate guidance may be issued.

*Satellite images*

Imagery both from radars and satellites makes a major contribution to this type of forecasting, as described by Booth (1984). The use of radar data will be discussed in some detail in the next section of this paper, and we concentrate here on the availability of satellite data.

Data from two types of meteorological satellite are available for use by the Met Office, as given in Table 2.1 and 2.2. The properties of the data streams from satellites are given in Table 2.3.

*Products derived from satellite data*

Images from the various satellites are available to forecasters in near real-time, and contribute to the appreciation of the weather situation and how it might develop. However, there is an increasing awareness that a number of value-added products can be derived from the raw data. These products include fog distribution

## Use of weather radar and satellite data in weather forecasting

TABLE 2.1. *Types of satellite data used by the Met Office*

*Near-polar orbiting satellites*
  The TIROS-N series of polar orbiters which carry instruments that provide;
  —high resolution images,
  —temperature and humidity profiles through the atmosphere,
  —data collection facilities.

*Geostationary satellites*
  Meteosat, the European geostationary satellite situated at the Greenwich meridian that provides;
  —images of the earth every thirty minutes,
  —data collection facilities.

  GOES-East, the US geostationary satellite situated near 76°W that provides, assuming the complete system is operational;'
  —analogue images from its own instruments,
  —analogue images from polar orbiting satellites,
  —analogue images from other geostationary satellites, including GOES-Central at about 108°W and GOES-West at about 135°W.

(Eyre *et al.*, 1984, Allam, 1987), sea surface temperature (Llewellyn-Jones *et al.*, 1984), land surface temperature (Browning, 1982) and snow cover (Bailey *et al.*, 1987). For example Fig. 2.1 shows snow cover over Scotland derived from AVHRR data. At present only images derived from analogue data are available to Met Office forecasters at outstations. This will change in the next year or so as a project known as Autosat-2 aims to provide digital imagery and derived products. This project will serve the needs of the Met Office and its customers for satellite imagery and data derived from it during the 1990s. As the name suggests, it will take over and expand the functions of the current operational system, Autosat-I.

The increased use of digital data streams, as opposed to analogue data, will result in finer spatial resolution, higher accuracy in the derived products, and the provision

TABLE 2.2. *Transmission formats of satellite data*

*Automatic Picture Transmission*—APT—analogue images from the polar-orbiting satellites;
*High Resolution Picture Transmission*—HRPT—digital data from polar-orbiters;
*Weather facsimile*—Wefax—analogue broadcasts of imagery from both geostationary and polar-orbiting systems;
*High Resolution*—HR—digital data from Meteosat.
(The digital S/VISSR data from GOES can be received in the UK. There is no requirement for them in the Met Office at present.)

Imagery is also available from the USSR "Meteor" near-polar orbiting satellites. "Meteor" imagery could be used in the event of the failure of the TIROS-N satellites. The USSR has indicated its intention to launch and operate a geostationary meteorological satellite (GOMS) at about 70°E. GOMS data would be of immediate interest to the Met Office.

TABLE 2.3. *The properties of satellite data-streams in 1988*

| Data-stream | APT | HRPT | Wefax | HR |
|---|---|---|---|---|
| Satellites supported | TIROS-N | TIROS-N | Meteosat, GOES (TIROS-N via GOES) | Meteosat |
| Type | Analogue | Digital | Analogue | Digital |
| Spatial resolution | 3 km | 1·1 km (best) 2·0 km (mean) | Meteosat<br>Vis— 2·5 × 2·5 km (best)<br>— 2·5 × 4·0 km (55°N)<br>IR/WV— 5·0 × 5·0 km (best)<br>— 5·0 × 8·0 km (55°N)<br>GOES<br>Vis (best) — 0·9 × 0·9 km<br>IR (best) — 6·9 × 6·9 km | |
| Data rate /kbit s$^{-1}$ | 33·3 (a) | 665·4 | 25·6 (b) | 167 |
| Daily data /Mbyte | 50 (a) | 1000 | 160 (b, c) | 660 |
| Radiometric resolution | 8 bit (0·4%) (d) | 10 bit (0·1%) | Meteosat<br>Vis/WV —6 bit (1·5%) (d)<br>IR —8 bit (0·4%) (d)<br>GOES<br>VIS/IR —8 bit (0·4%) (d) | |
| Equivalent precision of temperature from IR | (At 300 K) 0·35 K<br>(At 220 K) 0·92 K | 0·12 K<br><br>0·32 K | ←—— About 0·5 K ——→ | |

(a) Assumes an 8 bit sample at 4160 Hz
(b) Assumes an 8 bit sample at 3200 Hz
(c) Assumes 264 transmissions
(d) Degraded to 3–4 bits when displayed on facsimile paper.
NB: Both (a) and (b) are rates before conversion to analogue. Effective data rates will be smaller.

of a wide range of quantitative products to aid forecasts by providing animated sequences of pictures and for use in the Met Office's numerical models. Data from satellites will be received at RAE Lasham in Hampshire and processed on the Autosat-2 computers at that site. The processed products will be made available to users at outstations over the Weather Information Network (WIN) and to those in Bracknell over a dedicated communications link coupled to the Central Data

*Use of weather radar and satellite data in weather forecasting*

Fig. 2.1. Example of the use of AVHRR data from the NOAA polar orbiting satellites, to give snow cover over Scotland on 24 January 1986. White represents a high percentage snow cover and blue a low percentage snow cover. Cloud is shown as orange or black over land and green is land or shadows from cloud. (Courtesy of Dr E. C. Barrett, Remote Sensing Unit, Univ. of Bristol.)

Network. An important feature of the project is its interdependence with other major projects including the design of WIN, the outstation display system (ODS) and the Central Data Network in Bracknell.

*UK weather radar network*

The United Kingdom weather radar network was declared operational at the beginning of 1985. This was the culmination of many years of work centred on the Met Office Radar Research Laboratory (Met O RRL) at Malvern. Over the past few years the experimental network of radars has grown steadily, and data have been supplied to an increasing number of Met Office and water authority users. Building upon the software development carried out by a team from the Royal Signals and Radar Establishment, working with the Met O RRL in the early 1970s, the radar site and network software has been, and continues to be, developed by the Met Office for fully operational use. This work has provided a sound basis upon which further improvements in the radar site processing and in product availability will be carried out.

Currently there are six radars in the United Kingdom covering much of England, Wales and N. Ireland as shown in the inset to Fig. 2.2. All the radars are manufactured by Plessey Radar Ltd. Camborne and Upavon are S-band systems (10 cm wavelength, 2° beamwidth), and Clee Hill, Hameldon Hill and Chenies are C-band systems (5·6 cm wavelength, 1° beamwidth). The Hameldon Hill and Chenies installations contain modern Plessey 45C radars and are totally unmanned.

The most recent installation is that at Castor Bay in Northern Ireland which began operations towards the end of 1987. This radar was jointly financed by the Met Office and several other agencies in the Departments of Agriculture and Environment. The States of Jersey Meteorological Office hope to install a new digital radar, data from which will be included in the network. The new radars are to be sited at Preddanack in Cornwall and a site yet to be chosen in Dorset. The Preddanack radar should come into service in the spring of 1988, and will replace the Camborne radar. A further installation near Lincoln funded by the Met Office, Anglian Water Authority, Severn–Trent Water Authority and Yorkshire Water Authority is nearing completion. The network in England and Wales is to be completed by two further installations in south-west England and south Wales. Contracts have been placed for these systems which are to be funded by the Met Office, Wessex Water Authority, South-West Water Authority, Welsh Water Authority and Devon County Council.

Finally, a study recently completed by an *ad hoc* group led by the Met Office, but including representatives of the Scottish Development Department and a wide cross-section of industries, has identified substantial potential benefits from extending the weather radar network into Scotland. The benefits arise primarily from improved short-period weather forecasting, particularly for agriculture, road transport, and the building and construction industries. As a result the Scottish Development Department and Met Office hope to share the costs of a network of three radars in Scotland. These are likely to be located in the Outer Hebrides, to the

# Use of weather radar and satellite data in weather forecasting

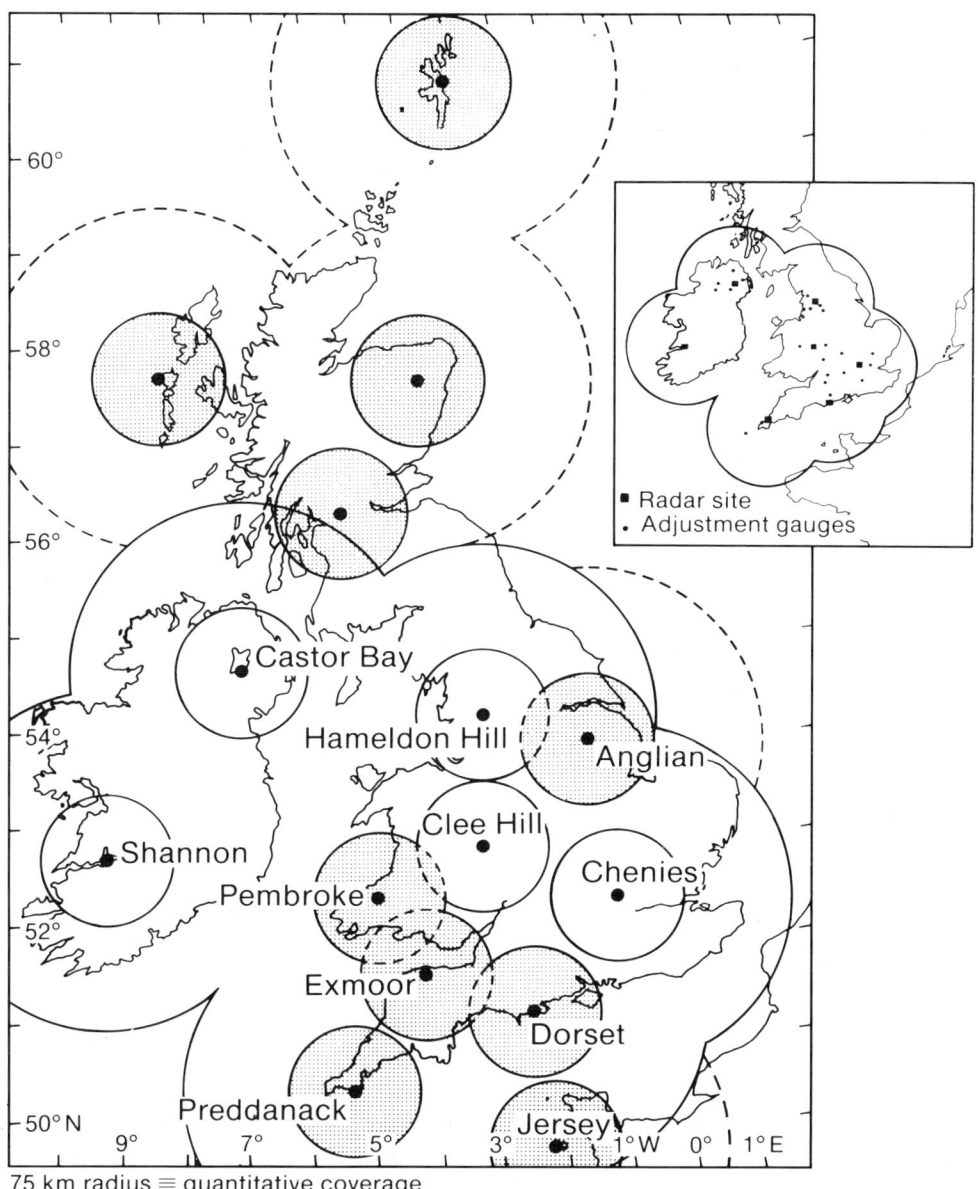

75 km radius ≡ quantitative coverage
200 km radius ≡ extreme range

FIG. 2.2. Location of existing and proposed weather radars in the United Kingdom. The inset shows the existing network (January 1988) and the location of raingauges used to adjust the radars in real-time. Planned radars (shaded) are shown at named sites in England and Wales.

north-west of Aberdeen and in the Central Lowlands. A fourth radar on the Shetland Islands would also benefit forecasting for oil industry activities there, but no plans exist at present to pursue this. Fig. 2.2 shows these possibilities.

(a) Radar data processing

Each radar site has its own dedicated minicomputer, either DEC PDP 11/40 or 11/34 systems. The software at most sites is written in CORAL-66, using DEC MACRO-11 for the time-critical areas, under the DEC operating system RSX-11M (or RSX-11S, a subset of this operating system). The software carries out the following main functions in real-time (Collier and James, 1986):
1. Removal as far as possible of ground echoes and correction for partial beam blockages by obstacles, mainly hills.
2. Conversion of received power to rainfall rate.
3. Identification of the presence and height of the bright band. (The region where snow melts to form rain produces an enhanced radar echo which is referred to as the 'bright band', see Smith, 1986).
4. Transfer from polar co-ordinates to Cartesian co-ordinates (2 km and 5 km grids).
5. Insertion of higher elevation data into the lowest elevation image to further remove ground echoes.
6. Adjustment using telemetering raingauge data (see Collier *et al.*, 1983).
7. Integration over river subcatchment areas.
8. Transmission to users and the radar network centre.

Some processing is carried out within the on-site hardware which for the newer installations includes an array processor which 'front ends' the DEC computer.

Each radar site transmits one picture every 15 minutes to the network centre at Bracknell where the data are processed by a dual DEC PDP 11/44 system. Synchronous communication links operating at 2400 baud provide the radar data to the system discs from which the network software constructs a composite image which is distributed to both forecasting offices and other non-meteorological users. The individual radar images are passed to the Frontiers system, described later.

(b) Accuracy of rainfall estimates by radar

Collier (1986) has indicated the accuracy of rainfall estimates that is achieved by implementing the real-time raingauge adjustment scheme, described by Collier *et al.*, (1983), which is in place at all the UK radar sites. Measurements of the energy backscattered from precipitation particles in volumes above the ground at many ranges out to 100 km or more, and at different azimuths as the radar beam rotates about a vertical axis, may be related to the rate of precipitation. Several sources of error have been identified (for a more detailed discussion see Browning, 1978):
1. Variations in the relationship between the backscattered energy and rainfall rate within the radar beam, due to dropsize distribution variations, the presence of hail in the radar beam, and also to the occasional presence of snow or melting snow. When the radar beam intersects the region where snow melts

to form rain, the radar reflectivity is enhanced producing the "bright-band". Measurements of radar reflectivity within the bright-band can lead to over-estimates of the surface rainfall.
2. Changes in the actual precipitation intensity, both within the radar beam and between the radar beam and the ground, due to raindrop growth or evaporation.
3. Variations in the performance of the radar system.
4. Attenuation of the radar signals due to heavy rainfall along the beam and due to the effects of water on the radome.
5. Radar signals produced by the ground especially when the path of the radar beam is affected by anomalous conditions in the lower parts of the atmosphere.

These errors are summarised in Fig. 2.3. The dominant importance of the first two types of error has been stressed by Browning (1981), although the others may contribute very significant errors on particular occasions. The accuracy of the radar measurements of rainfall may be improved somewhat if a raingauge is used to adjust the system, as shown by Wilson (1970).

An empirically derived range-dependent correction profile is applied in real-time to the radar to allow in part for increasing under-estimation of rainfall at far ranges, especially in winter, due to incomplete filling of the radar beam or to failure of the radar to observe low-level rainfall. This correction profile was derived by comparing hourly estimates of rainfall made using the radar with raingauge estimates at various ranges from the radar over a number of rainfall events associated with different

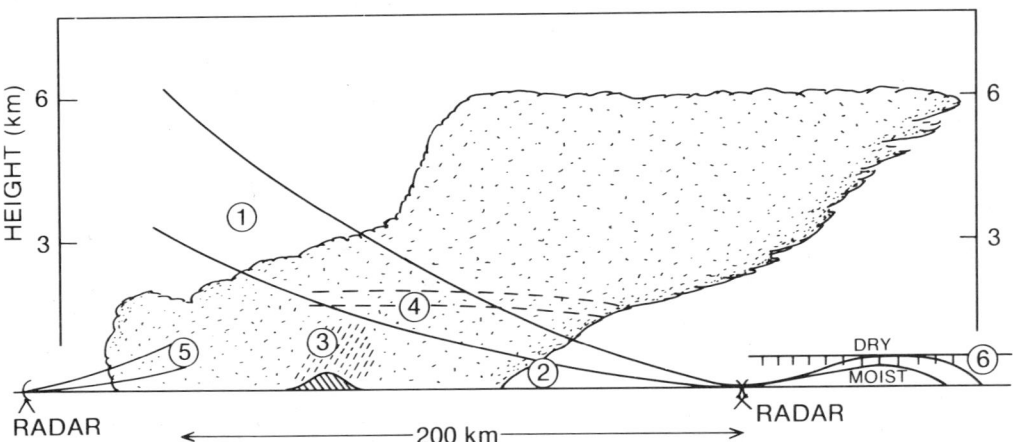

KEY: (1) radar beam overshooting the shallow precipitation at long ranges, (2) low-level evaporation beneath the radar beam, (3) orographic enhancement above hills which goes undetected beneath (4) the bright-band, (5) underestimation of the intensity of drizzle because of the absence of large droplets, and (6) radar beam bent in the presence of a strong hydrolapse causing it to intercept land or sea.

FIG. 2.3. Cross-section through an area of frontal precipitation illustrating six sources of error in the radar measurement of surface rainfall intensity (from Browning, 1981).

weather types. A tendency to over-estimate at near ranges is also evident. The over-estimation may be due to the presence of the bright-band in the radar beam at close ranges, while the interception of snow and a lack of detection of low-level orographic rainfall enhancement probably causes the under-estimation at far ranges. During the summer radar gives good estimates of rainfall to quite long ranges.

At any time of the year the individual assessment factors may vary considerably from the mean monthly values. The average percentage difference (regardless of sign) between radar estimates of hourly point widespread frontal rainfall and gauge estimates is 45% for gauge adjusted data and 60% for unadjusted data, provided no bright-band effects are present. For convective rainfall the average percentage difference between radar estimates of hourly point rainfall and gauge estimates is 21% and 37% respectively. For data with bright-band effects the average percentage difference between radar estimates of hourly point rainfall and gauge estimates is 75% for adjusted data and 100% for unadjusted data.

Radar is also capable of making measurements of snowfall as accurately as those of rainfall except for greater wind drift problems beneath the beam. Studies over flat terrain have been reported by Jatila (1973) and Pollock and Wilson (1972), and Collier and Larke (1978) have shown that comparable results can be obtained over hilly terrain in N. Wales.

*The Frontiers[1] system*

The Frontiers system is a computer-based system for merging radar and satellite data, and eventually other meteorological data, to produce precipitation nowcasts, i.e. detailed descriptions of the current distribution of precipitation plus forecasts for a few hours ahead, obtained by extrapolation. The system was designed and built by Logica Ltd to a Met Office functional specification, and delivered in initial form in 1983 (Browning, 1979, Carpenter and Browning, 1984). Since then considerable effort has been put into bringing the system to operational status by September 1986 in the Central Forecast Office at Bracknell.

(a) System structure

The system operates on a half-hourly cycle synchronised with Meteosat's tranmissions. The cycle is divided into three major stages (Conway, 1987). On-site range corrections and raingauge adjustment are removed as the data enter the system.

In the radar analysis stage the operator aims to produce the best possible estimate of the intensity and areal extent of the rainfall within the area covered by the radars. This is achieved by using data from other sources (conventional observations, Meteosat cloud-images, etc), and his understanding of the meteorological situation to help recognise and correct errors in the radar data. Detailed maps of orographic enhancement are applied depending upon low level wind velocity and humidity

---

[1] Frontiers—Forecasting Rain Optimised Using New Techniques of Interactively Enhanced Radar and Satellite data.

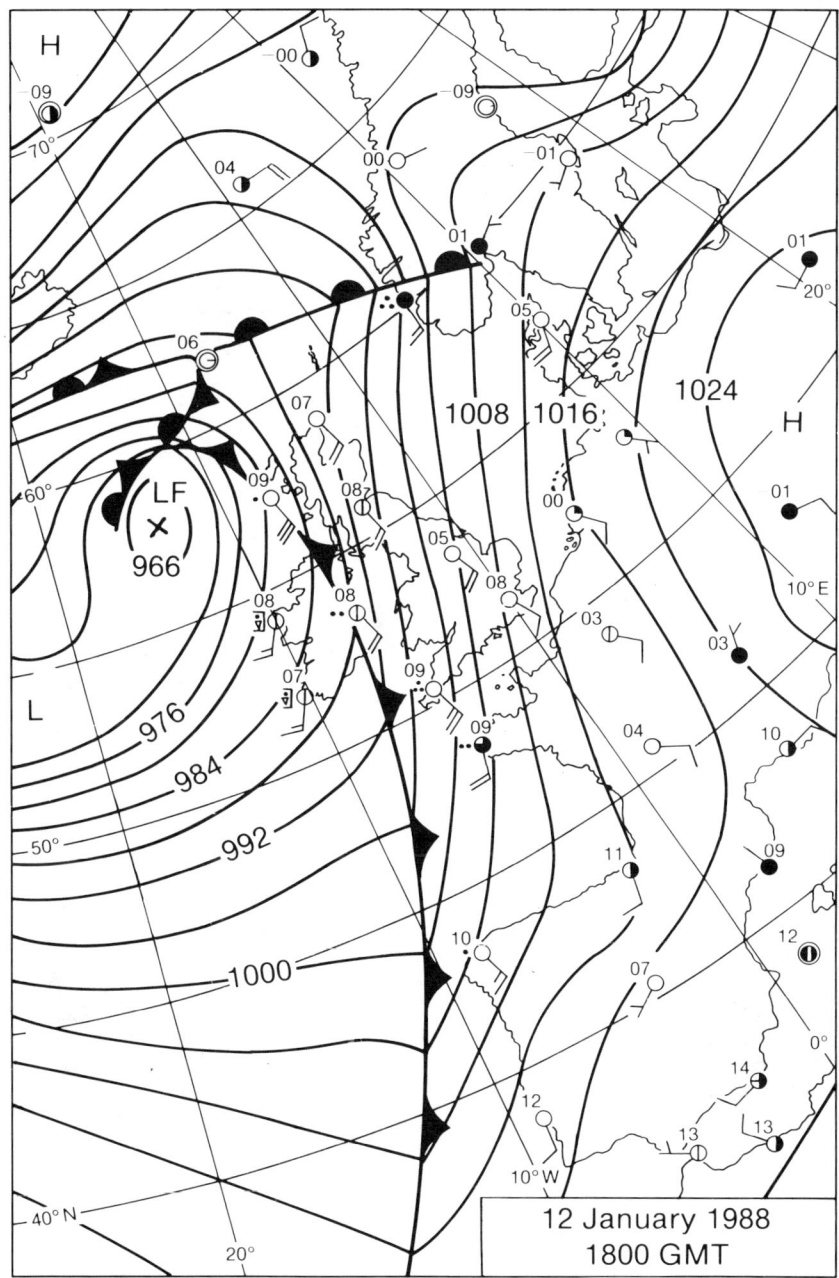

FIG. 2.4. (a) The synoptic situation over the UK at 1800 GMT on 12 January 1988; (b) Forecast combined radar and satellite images produced operationally at 1400 GMT on 12 January 1988 for 6 h ahead with the corresponding verification radar images. Satellite data are shown brown, other colours representing rainfall intensities derived from radar data. Note the success in forecasting the rainfall clearance over Ireland.

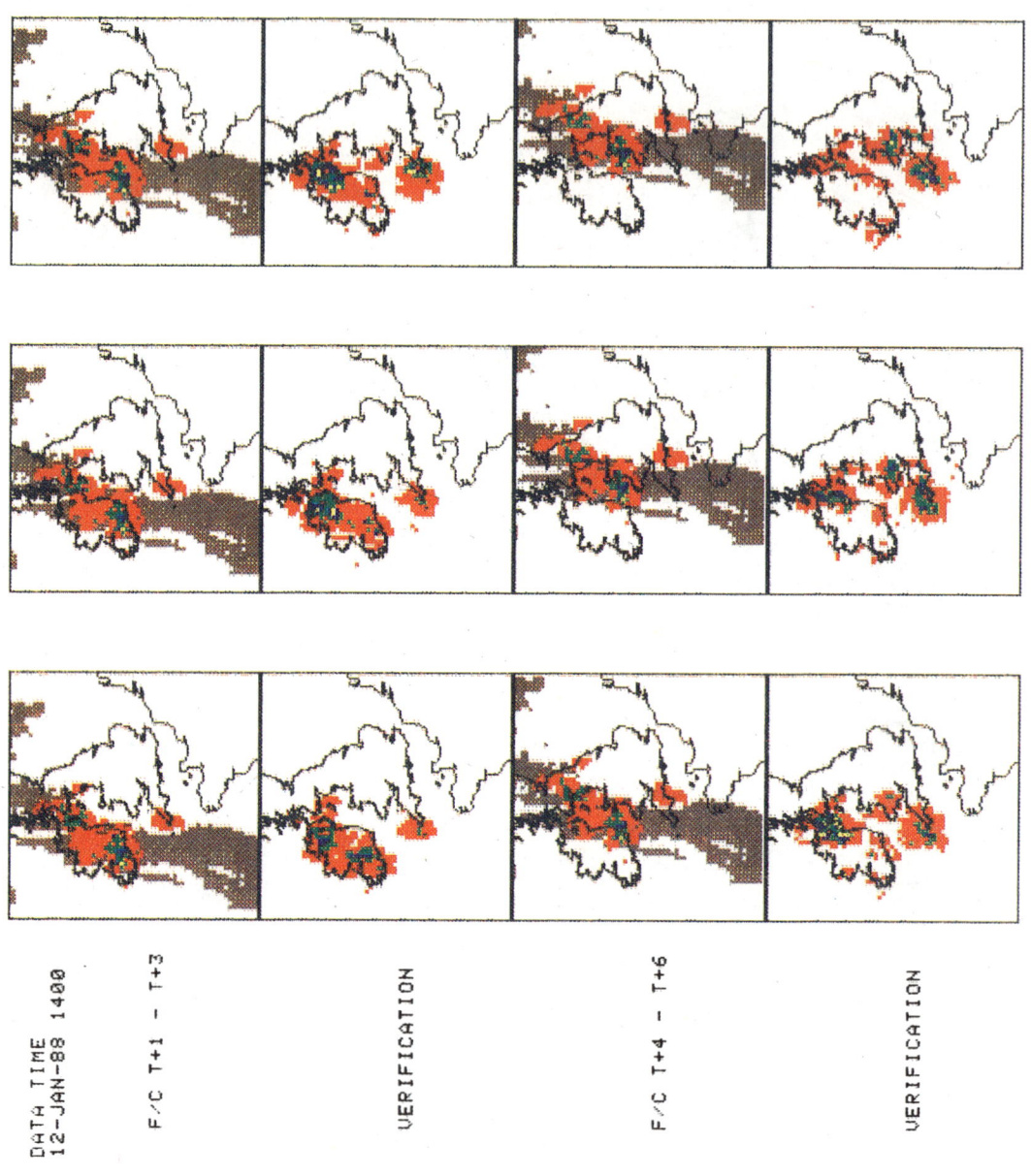

FIG. 2.4 (*continued*)

specified by the operator. Finally the data are re-adjusted using raingauge data. The raingauge adjustment step is presently being re-designed to be more compatible with the orographic enhancement and remote site procedures.

In the second stage the operator extends the area of coverage beyond the radar area by attempting to deduce the extent (but not the intensity) of rainfall over the full Frontiers area from Meteosat cloud-images. The satellite-derived rainfall is merged with the quantitative radar rainfall map from the radar analysis stage to provide as complete a picture as possible of current rainfall in the Frontiers area.

The final stage is the production, by the linear extrapolation of recent movement, of a sequence of rainfall forecasts covering the next six hours. The operator divides the rainfall field from the second stage into a number of clusters to which are then assigned independent velocities. Frontiers then computes the forecast by moving the clusters with their assigned velocities.

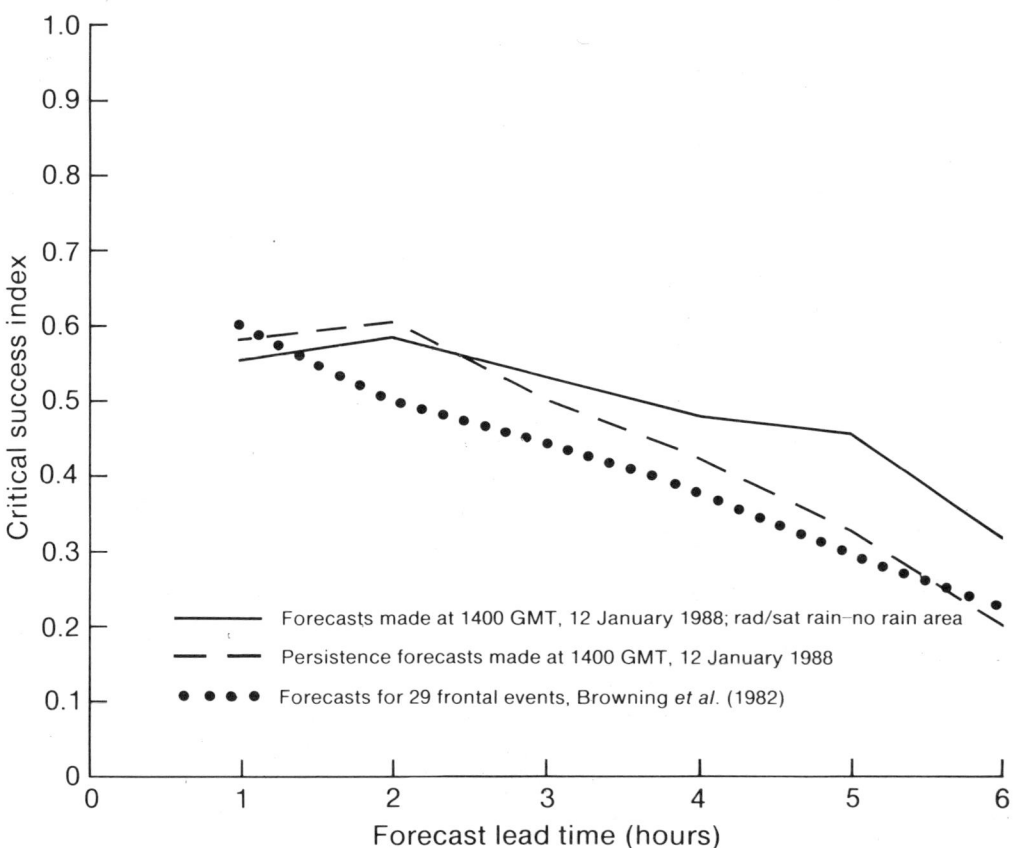

FIG. 2.5. Variation of the Critical Success Index (CSI), with forecast lead time for rain-no rain forecasts made operationally at 1400 GMT on 12 January 1988.

(b) Forecast accuracy

At present the forecasts are produced with a 20 km resolution, but work is nearing completion to change this to 5 km, compatible with the input data. Examples of the radar and satellite actuals with forecasts for up to six hours ahead are shown in Fig. 2.4. A full assessment of the current Frontiers forecasts is not yet completed.

An indication of the performance is given in Fig. 2.5. in which an objective forecast score is shown as a function of lead time for the case given in Fig. 2.4. For comparison, earlier off-line work for frontal rainfall reported by Browning *et al.* (1982) is also shown in Fig. 2.5. The assessment is just for rain-no rain areas over the area of best radar coverage. Note that the Frontiers forecasts are significantly better than persistence as the lead time is extended. Much more work is needed before the forecast performance can be specified as a function of synoptic weather type. In order to identify potential problem areas, a trial dissemination of the forecasts to the North West Water Authority at Warrington started at the beginning of 1988.

*Nowcasting*

The production of detailed and timely forecasts for a few hours ahead tailored to the needs of specialist customers requires detailed analysis of both radar and satellite imagery, as shown in the previous section. This type of forecast is known as a nowcast as it is based upon detailed knowledge of the weather at the time the forecast is made. Systems such as Frontiers have demonstrated the feasibility and potential of blending radar, satellite, numerical model, and conventional data in near real-time. In the UK the basis of this work is provided by data from the network of weather radars, and is complementary to numerical forecasting. Indeed the preparation of nowcasts does require the consideration of numerical products.

## USE OF RADAR AND SATELLITE DATA IN NUMERICAL WEATHER FORECASTING

Forecasts produced using nowcasting techniques depend upon manipulation, including extrapolation, of basic data. No numerical integrations of the equations governing atmospheric motions are carried out, although, as mentioned above, nowcasts may use numerical products derived from other sources. Most routine operational numerical forecasting uses models which provide valuable guidance to forecasters on the broad scale atmospheric structure. However, these models have large (>75 km) grid lengths, and therefore cannot properly represent detailed topographic differences which are important for short-period forecasting, and other mesoscale systems unrelated to topography. Hence models are being developed (Tapp and White 1978, Carpenter 1979, Pielke, 1981) with much smaller grid lengths (15 km), and much more detailed physical parametrisations of processes occurring on scales smaller than the grid length. These models are known as mesoscale models. A further class of models attempts to forecast the movement of pollutants, including radioactivity. These models too will be discussed briefly.

## Mesoscale models

Because mesoscale model grid lengths are small, and physical parametrizations complex, the initial data used in integrations must be specified in real-time accurately, otherwise the model will deviate rapidly from the observed atmospheric evolution. Likewise, model boundary conditions must also be specified accurately, particularly the lower boundary condition. Zhang and Fritsch (1986) and Golding (1987) emphasise the importance of both topographic and thermal forcing to the accuracy of predictions, and note that correct representation of the responses to such forcing requires the accurate specification of the moisture distribution in the atmosphere. This is because the primary response of the atmosphere on these scales appears to occur through condensation of water, and the consequent release of latent heat. Support for this comes from numerical experiments reported by Diallo and Frank (1986). Thus, a detailed specification of the three dimensional moisture distribution is required. Given this, the distribution of condensation and consequent latent heating is determined, and applied over a period of integration at the start of the forecast. This moisture distribution may be estimated from satellite and radar data.

## Operational modelling for short- and medium-range forecasting

Most regional operational numerical forecast models, for example, that of Bell and Dickinson (1987), aim to produce weather forecasts covering the period 1–5 days ahead. The models use grid lengths of 50–150 km, and are applied to restricted areas relevant to weather forecasting for the area of interest. Beyond these lead times it is essential to use global models (for example, Burridge, 1979; Bengtsson, 1985; Gadd, 1985) as weather events many thousands of kilometres remote from the area of interest affect the evolution of atmospheric systems which ultimately produce the weather in these locations.

In many respects, the structure of these models resembles that of mesoscale models as the same equations are used. However, the coarse resolution makes it necessary to parameterise physical processes, particularly those associated with small scale energy exchanges within the boundary layer and within clouds. Nevertheless, over the last decade there has been a steady improvement in forecast skill, such that forecasts for three days ahead are now as accurate as the two-day forecast was in 1978 (Flood, 1985). This improvement in forecasting skill is also evident in the number of days ahead that forecasts can be made with some confidence of success, now over six days ahead (Bengtsson, 1985).

In spite of these advances, many problems remain, including the lack of wide coverage information from which to derive humidity and latent heat release. Rainfall forecasts for the United Kingdom have only improved slowly, but have reached a level which is operationally useful. In general, heavy rainfall is not forecast well, and the model tends to over-predict light rainfall. The finer grid length of the mesoscale model produces an improvement, but forecast accuracy is still not uniformly high. It must be clearly understood that the mesoscale model is still under development, and the present performance is very much a first step. The use of satellite and radar data in the way described is expected to improve forecast accuracy. Experiments are underway to assess how much improvement can be achieved (Golding, 1988).

*Modelling atmospheric dispersion and deposition*

Following the nuclear accident at Chernobyl, thought has been given in the UK as elsewhere to means of improving the national response in the event of any further contingency of this kind. The Met Office are the lead agency in a plan to develop an analysis and forecast trajectory model for the dispersion and deposition of radioactive material.

Roesli *et al.* (1987) in Switzerland and ApSimon *et al.* (1988) in the UK have shown the close correspondence between rainfall as depicted by radar, and radioactive material deposition during the Chernobyl event. The need to have available in real-time information on precipitation extent and amount has been recognised by Smith (1987) in the design of the UK model.

*Wide area precipitation analysis package*

The need of numerical models of various types for wide area precipitation information has been discussed above. No one source of data can provide the information required. Whilst radar can provide acceptable accuracy for many applications the coverage available is rather restricted. This is in spite of work such as the COST-73 Project which aims to coordinate the international exchange of radar data throughout much of Europe (Collier *et al.*, 1988). Nevertheless, radar data in combination with geostationary satellite data, numerical model data, and conventional observations are likely to satisfy the requirement. Work is underway in the Met Office to develop a software package which will provide hourly fields of precipitation rate and type over the areas needed by the mesoscale and trajectory models.

*Soundings of temperature and humidity*

The HIRS (High resolution Infra-red Radiation Sounder) and MSU (Microwave Sounding Unit) instruments on the TIROS-N series of polar orbiting satellites provide data which may be processed to give temperature and crude humidity soundings through the atmosphere. NOAA/NESDIS in Washington DC provide soundings which are distributed internationally as so called SATEM messages (Smith *et al.*, 1979). These soundings have a resolution of 250 km and are used in the UK as elsewhere in the operational analysis scheme which provide input data to the numerical models.

Soundings may be derived at a higher resolution than the SATEMS and may be obtained more quickly if the data are received directly from the satellites and processed locally. Over the last few years the Met Office have developed the HERMES[1] system (Eyre and Jerrett, 1982). The retrieval scheme in current use

---

[1] HERMES — High-resolution Evolution of Radiances from MEteorological Satellites.

was obtained from the NOAA/NESDIS Development Laboratory at the Cooperative Institute for Meteorological Satellite Studies (CIMSS), Madison, Wisconsin. Soundings are derived in cloud free areas, and used to derive 1000–500 hpa thickness fields and thermal winds. Currently several improvements are being assessed and impact studies on the numerical analysis are being carried out. Problems remain in removing inconsistencies between satellite swaths causing errors which amplify rapidly in the numerical forecasts. However, significant positive impact has already been demonstrated in the N. Atlantic on some cases, and most of the time in the data sparse areas of the southern oceans (Bengtsson, 1985).

Present research is concentrating on using the numerical forecast as a first guess instead of climatology to define the thickness and thermal wind fields. Iteration, using observed temperature profiles produces an improvement on the original field. This technique will also be exploited in developing a retrieval scheme to process the data from the new instruments to fly in a few years time such as the Advanced Microwave Sounding Unit (AMSU). The Met Office are contributing to this instrumentation by developing with industry the AMSU-B instrument which will use radiometers which operate at 89, 157 and 183 GHz (Pick, 1986).

## CONCLUSION

This paper has outlined ways in which radar and satellite data are, or soon will be, contributing to operational weather forecasting. Much of what has been discussed demonstrates the complementary nature of radar and satellite data. Observing systems invariably require both types of data to satisfy the requirements of weather forecasting.

It is already evident that significant improvements will be achieved in forecast accuracy as a direct result of the use of remotely sensed data. New satellite systems already being planned will undoubtedly provide further useful data. However, it will be important in the future to devote appropriate resources to ground segment processing, and systems devoted to blending data from widely differing sources. Finally, in the area of remote sensing it is often easy to overlook the actual requirements of users. Care will have to be taken to focus research and development in ways which will lead to improved products and services. This does not mean that fundamental research should not be carried out, but more that sparse resources should be carefully husbanded. It will be important to recognise those areas of research which are likely to underpin developments in operational services.

ACKNOWLEDGEMENTS. The help of members of the Satellite Meteorology Branch (Met O 19) of the Meteorological Office is acknowledged. In particular R. Allam, R. Brown, Dr. B. Conway, D. Goddard, and G. Sargent have provided material from their research work for this paper. Thanks are also due to Dr E. Barrett, Univ. of Bristol for the use of Fig. 2.1.

# The use of numerical models in weather forecasting: achievements and prospects

### B GOLDING

*Meteorological Office, Bracknell*

## BACKGROUND

The aim of weather forecasting is to describe the weather for a location or area for some period ahead. In order to do this scientifically, information about the atmosphere must be organised into a coherent structure. Traditionally, the meteorologist has used the sea-level atmospheric pressure distribution as his basis, adding fronts to indicate boundaries between air with different characteristics, usually accompanied by cloud and rain. For a period of a day or so ahead, the experienced forecaster will use his knowledge of both surface and upper air patterns to predict how this representation of the atmosphere will change, and then interpret his prognosis in terms of expected wind, temperature and rainfall. Note here the distinction between prediction of general atmospheric development, and interpretation of specific weather elements. For longer periods ahead a better structure to predict is the pattern of the jet streams high above the surface. The diagnosis of weather at the surface is in this case more generalised. However, in both approaches the interpretation is basically statistical, relying on past occurrences of this type of air mass or boundary in the location of interest. For short period forecasts, up to a day ahead, the user of weather forecasts will often need to know more precisely how far inland cloud will penetrate, whether it will produce rain, whether the rain will turn to snow, or whether fog will clear. Such forecasts can only be made with detailed knowledge of the air involved, taking account of changes as it travels over sea and land of differing surface characteristics. Allowance must also be made for feedback mechanisms which may operate in the atmosphere, and on the influence of the changing large scale atmospheric structure. At the very short range the problem becomes one of deducing the current state of the atmosphere from limited and imperfect observations.

The development of scientifically based weather forecasting proceeded from short to longer periods ahead, as one would expect. Starting from the realisation that communication of one person's observations to another may forewarn the second of adverse weather to come, it proceeded to the recognition of patterns in the surface weather observations which could be organised with reference to the pressure pattern, and finally to the use of the upper air motion pattern to determine how the surface pressure pattern will evolve.

In some respects the numerical modelling method has developed in the reverse direction. The pattern of upper air troughs and ridges can be described at a coarser

## The use of numerical models in weather forecasting: achievements and prospects

resolution than the surface pressure pattern. Also, the physical laws required to give a usable description of the upper air evolution are far simpler than those needed to represent surface weather. Thus, early numerical models predicted only the upper air pattern (Bushby, 1987) and the forecaster inferred the surface weather from it. However, not only could this lead to wrong inferences at the surface, but the poor representation of the lowest layers of the atmosphere could affect the overall development on some occasions. Later, increased computer power permitted greater complexity with enhanced resolution in both the vertical and horizontal (Burridge and Gadd, 1977). Such models could predict the surface pressure pattern and main rain belts fairly reliably. Since that time, further increases in power have allowed prediction of more detail while additional variables have been made available to the forecaster to assist his interpretation of, for example, surface winds, humidity and stability. We are now on the threshold of implementing models which have the potential to directly predict almost all of the weather elements required by users. Again, this has been made possible by increases in computing power which permit more complex models with better horizontal and vertical resolution.

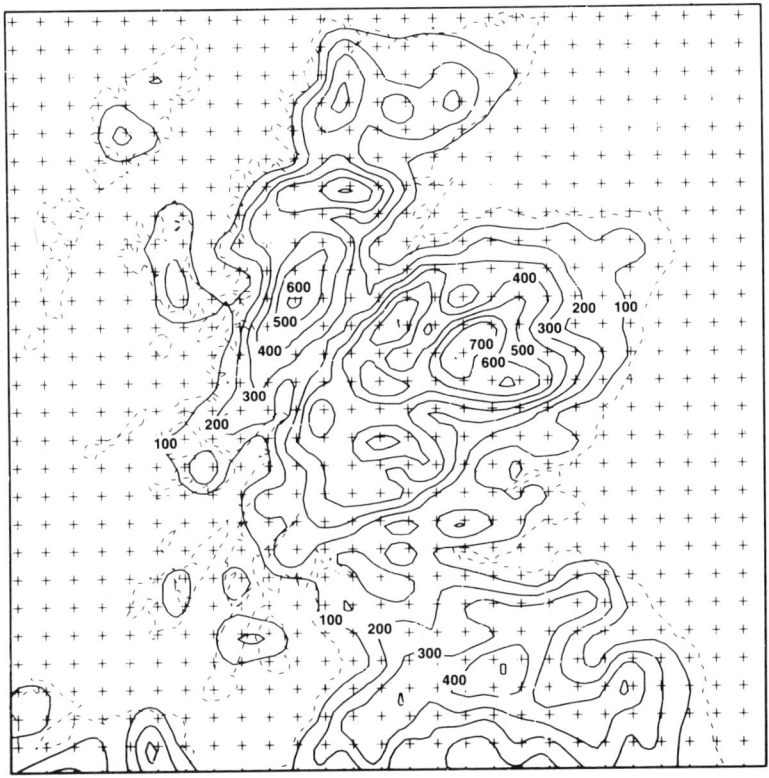

FIG. 3.1. Grid points and surface elevation contours of the experimental mesoscale model. Contour interval: 100 m.

To illustrate the requirement, consider the representation of the land surface which is crucial for many elements, particularly temperature. Whereas a resolution of 100 km gives a realistic description of the pressure map on most occasions, the 10 or so grid points representing Scotland make no pretence of portraying the detailed pattern of surface elevation. With the 15 km resolution model being tried experimentally at the UK Met Office (Golding, 1987), the description appears realistic (Fig. 3.1.), but is still incapable of resolving important detail in the hillier parts of the country.

## CURRENT CAPABILITIES

Beyond about two days ahead, it is necessary to take account of the state of the atmosphere over the whole globe in order to predict the changes in any particular place. Such models are run operationally at a small number of centres including Bracknell, Washington and the European Centre for Medium Range Weather Forecasting (ECMWF). Since minor differences near the start of a forecast can sometimes become greatly magnified over a long period, it is beneficial to have several predictions to compare. The forecasts issued by the Central Forecast Office in Bracknell are, therefore, based on a critical review of model predictions from all of these centres. The global model run at Bracknell (Bell and Dickinson, 1987) currently has grid points spaced about 150 km apart in this part of the world, and represents the atmosphere by 15 layers in the vertical. As well as modelling the fundamental laws governing the motion of air on a spinning globe, this model includes the effects of moisture in producing clouds and rain, heating and cooling due to radiation transfer, and the mixing processes which spread heat and moisture in the near surface layers. This model, and others like it, produce useful forecasts of the upper air pattern out to 4 or 5 days ahead on average, though sometimes it can be as long as 10 days, and sometimes as little as a day.

At this range, a useful forecast is one which identifies the general character of the weather over the British Isles, not attempting to locate specific rain belts or storms. For this reason, forecasts for the general public for this far ahead must be worded very carefully to cover the possible weather that may be associated with a reasonable degree of uncertainty in the numerical predictions. It should be noted that forecasts to this range were not practicable without numerical models and that recent advances have given five-day forecasts the accuracy that two-day ones had in the early 1970s. Given specific requirements by an individual customer such as for a weather window for an offshore operation, the forecaster may often be able to be more specific in his forecast even at this range.

An example of model performance is shown in Fig. 3.2. The verifying surface analysis in (a) shows a vigorous depression to the west of Ireland with rain-bearing fronts across western parts of the British Isles. At six days ahead (Fig. 3.2b) the forecast pattern bears only a weak resemblance to the actual. Nevertheless, it gives a general picture of disturbed conditions with winds and rain belts over the British Isles. Clearly even this would be sufficient to rule out a weather window in the

## The use of numerical models in weather forecasting: achievements and prospects

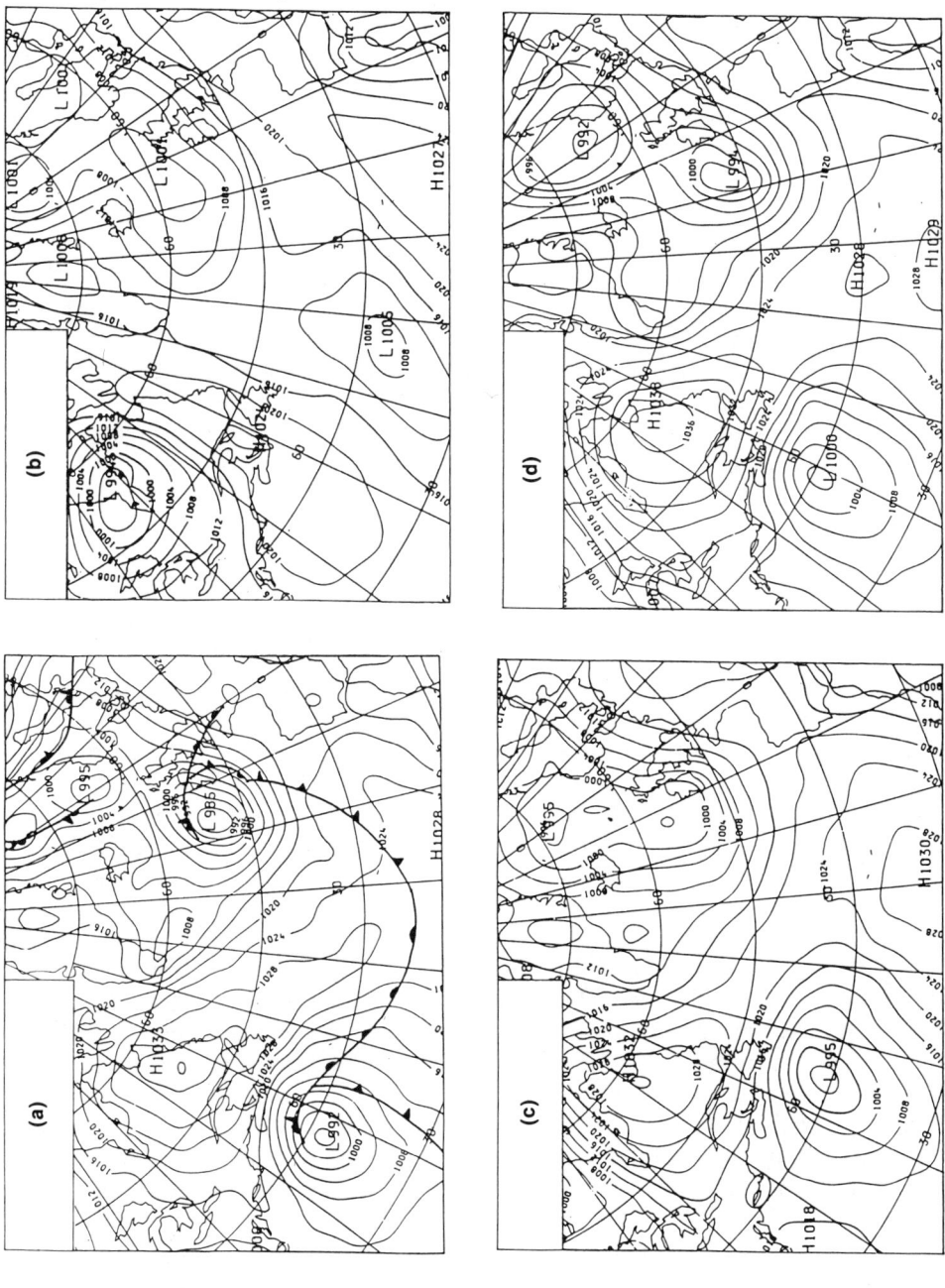

FIG. 3.2. North Atlantic surface charts for 0000 GMT on 12 May 1986. (a) analysis with hand drawn fronts (b) 6-day forecast (c) 4-day forecast (d) 2-day forecast.

North Sea. The four day forecast (c) does significantly better in locating the depression near Ireland and at this stage the consistency between successive runs would give the forecaster some confidence in the general evolution. Finally in Fig. 3.2d, the two day forecast was on this occasion a very good representation of the actual pressure pattern. Note, however, that the implied position of the forward rain belt in this forecast is southwest of the actual, giving a timing error which might be considered significant in a short period forecast.

For a day ahead, the contribution of numerical models has gradually increased over 20 years or more so it is difficult to detect the improvements. In fact, the predictions are very good at this range when assessed against a tolerance of, say, 6 hours in passage of a rain belt. The exceptions usually come with small rapidly developing systems which may be missed in their early stages by the sparse oceanic observations, and whose development and movement is particularly sensitive to the details of the atmospheric circulation in which they are embedded. These also tend to be the more violent systems so the elimination of such errors is of particular importance. In order to model such detail and to give guidance on showers, and cloud cover between the main rain belts, limited area models are run with a finer resolution than the global models. Several European Meteorological Services run such models with a resolution better than 100 km, centred on their own geographical areas. At Bracknell, a model with 75 km resolution is run to 36 hours ahead, twice each day. it covers much of the North Atlantic and Europe with most interest in the predictions for the British Isles and surrounding waters. As well as the pressure pattern, this model gives useful predictions of rainfall, and away from mountains and coasts it can give a good estimate of surface temperature and wind. These are further improved for specific sites by applying statistical regression equations based on a comparison of past forecasts and observations. Over the sea, wind predictions are of very high quality and are used as input to numerical models of waves (Golding, 1983) and surges (Flather, 1981). These are of particular use to coastal defence organisations and the offshore industry. Development of these products has benefited from the close relationship between the forecaster, who is in touch with the customer, and the numerical modelling research groups in the Met Office organisation.

At shorter forecast periods, these models are not less accurate but the requirements are more detailed. Distinction is required between the coast and inland and persistence or clearance of fog and frost. At present, the operational provision of such information is based largely on subjective forecasting techniques within the framework of the large scale evolution up to a day ahead. Some parameters are reliably forecast with these techniques but there are considerable difficulties with cloud and visibility. It is to give guidance on such problems that the experimental mesoscale numerical model has been developed at Bracknell (Tapp and White, 1976; Carpenter, 1979; Golding, 1987) and is now being assessed both centrally and at a selection of forecasting sites around Britain, including Glasgow Weather Centre and the airfields at Leuchars and Lossiemouth on the east coast.

It has a 15 km resolution and represents the atmosphere in 16 layers, the lowest at

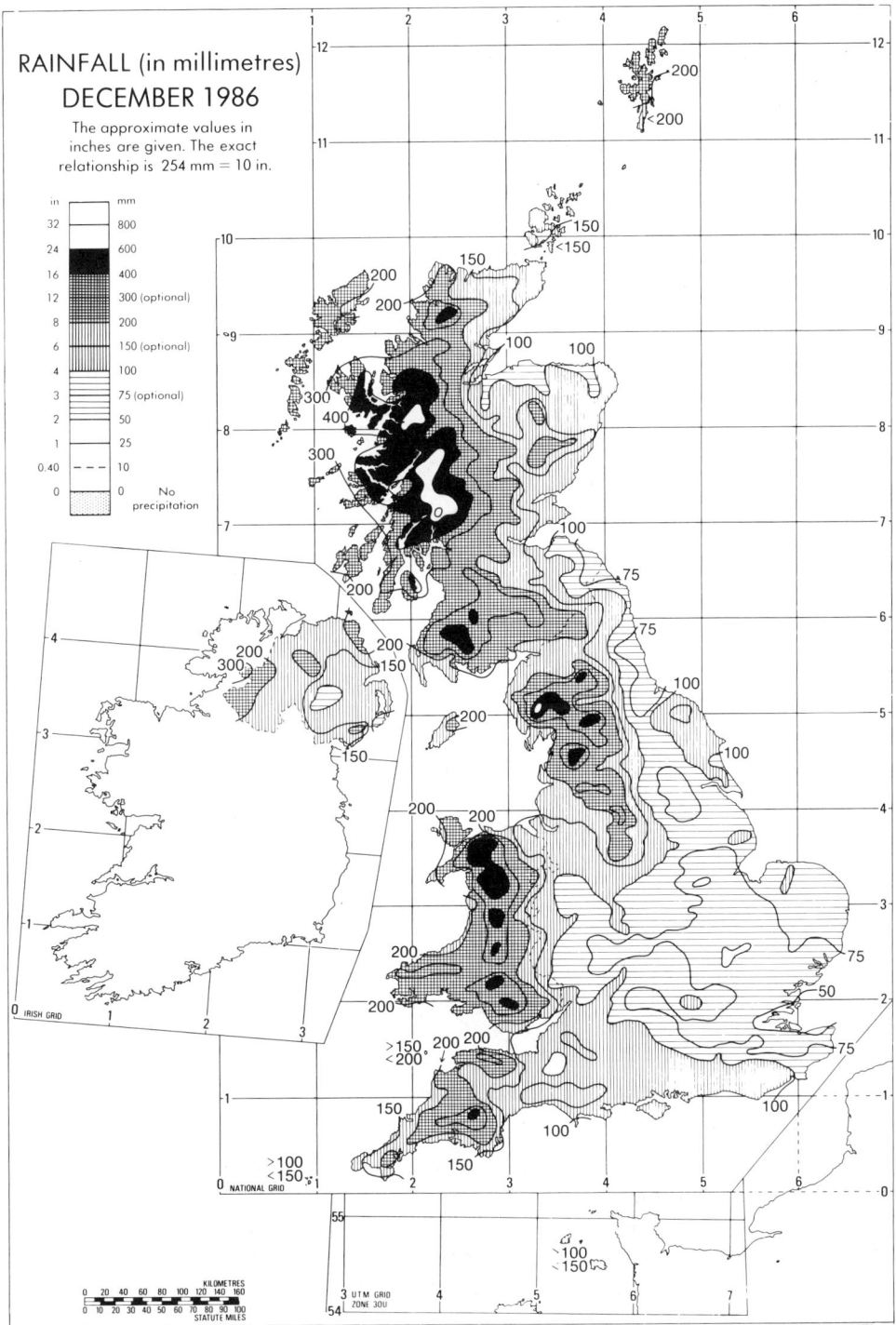

Fig. 3.3. Total precipitation in December 1986 (a) deduced from a dense network of raingauges (b) accumulated from successive 12-hour predictions.

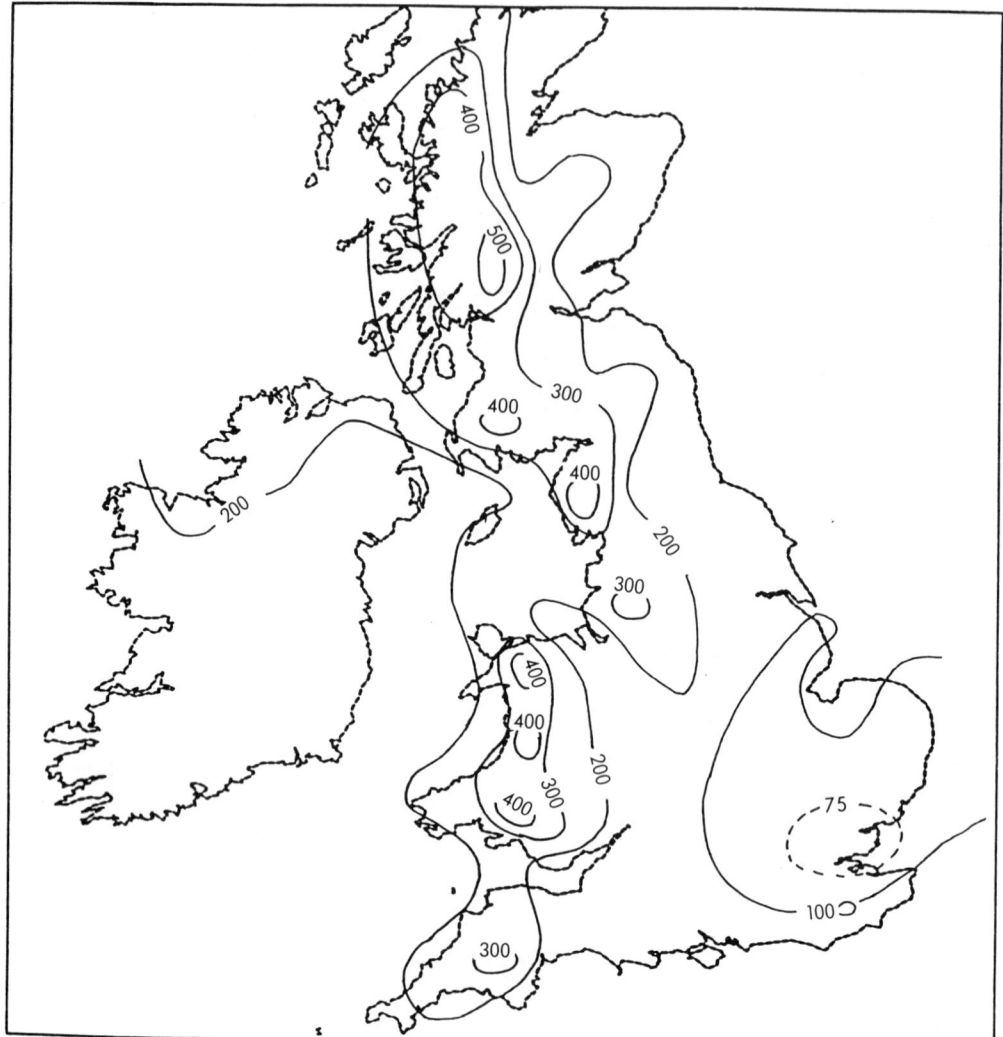

Fig. 3.3. (*continued*)

10 m above ground level. The results are, in general, encouraging especially for wind and temperature forecasts. Over the Highlands, however, the results are difficult to interpret because of the roughness of the real terrain compared with that in the model (Fig. 3.1). The accuracy of precipitation forecasts is difficult to measure because of natural variability associated with showers. However, a general view of the model's ability in this area can be obtained from monthly totals of successive twelve-hour forecast periods. Fig. 3.3 shows such a comparison for December 1986. The high values over the west of Scotland, Wales and England compare well with

*The use of numerical models in weather forecasting: achievements and prospects*

observations which here use the dense climatological raingauge network. In the east the model has overpredicted the smaller observed values a little.

Figs. 3.4 and 3.5 show an example of a particular case on 13 January 1987. The prevailing airstream was a very cold easterly and many places had a maximum temperature below −5°C on the 12th (Brugge, 1987). With sea temperatures of 8–10°C there was considerable convective mixing resulting in cloud formation and snowfall (Monk, 1987). Fig. 3.4a shows the shower distribution observed by the radar network at 0300 GMT and Fig. 3.4b the model's nine-hour forecast for this time. Over successive runs, the model accumulates snow depths giving evidence of

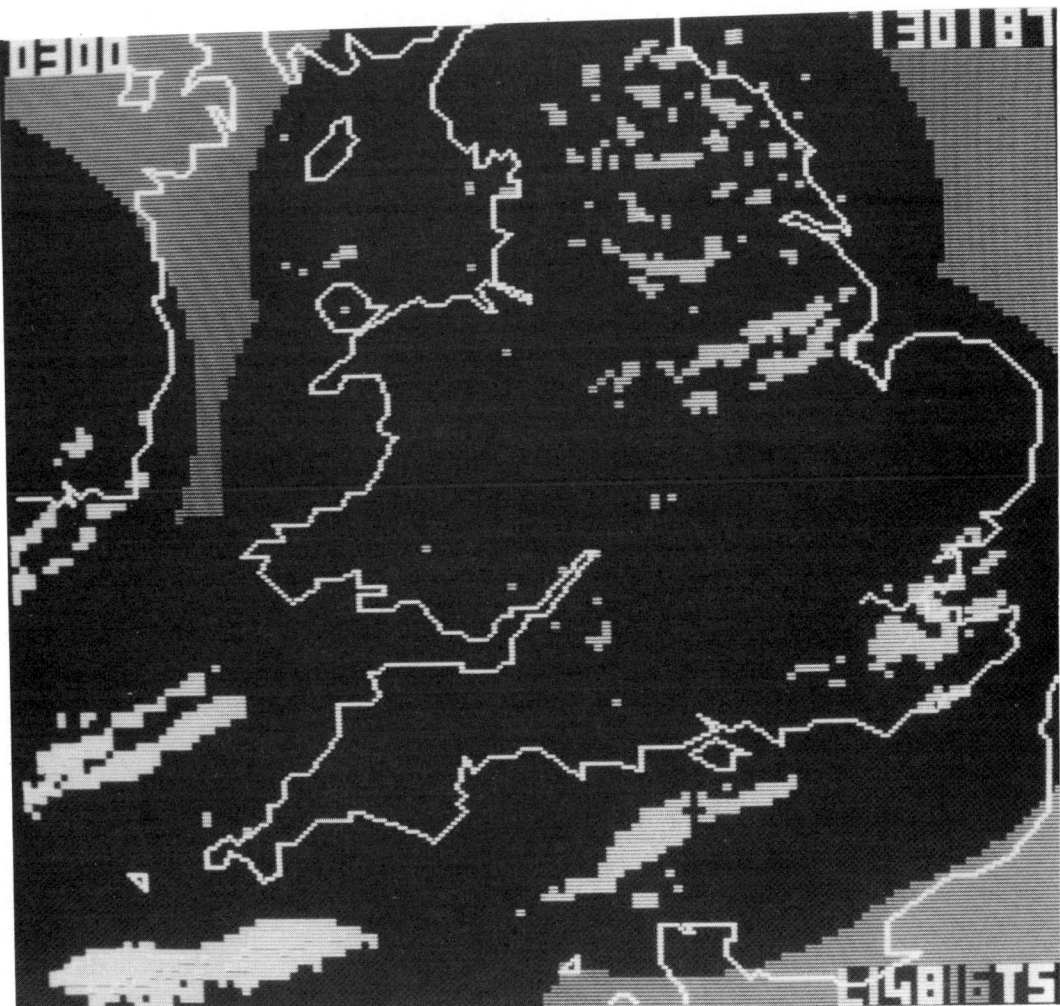

FIG. 3.4. Distribution of snow showers at 0300 GMT on 13 January 1987 (a) composite radar picture based on 6 radars. (b) 9-hour model forecast.

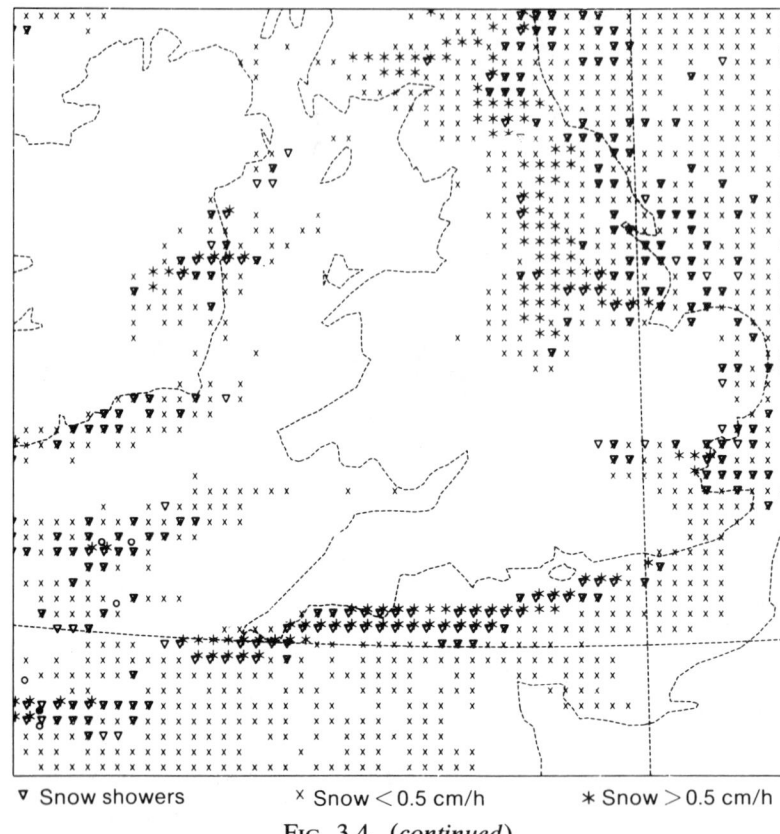

▽ Snow showers     × Snow < 0.5 cm/h     * Snow > 0.5 cm/h

FIG. 3.4. (*continued*)

the accuracy of its predictions over the period. In Fig. 3.5 the comparison is made with observations at 9 am on 13 January. Again the comparison is good, especially in the Thames Estuary and in Cornwall.

Turning to the difficult problems of cloud and visibility predictions, Fig. 3.6 shows a series of forecasts of cloud base at Leuchars in eastern Scotland during a period of persistent low cloud in May 1987. Three-hourly observations are plotted at the nearest model level to allow a fair comparison. Over this period, the only serious error was in the pessimistic forecast from midday 24th. A particular forecasting problem in northeast Scotland is the sea fog or haar. A field experiment was conducted in the Moray Firth in Spring 1984 using the Met Office's instrumented C130 aircraft and balloon ascents (Findlater *et al.*, 1988). In late April, a fog bank formed west of Ireland and moved around the coast of the British Isles until by the 27th it had filled much of the North Sea, affecting coasts of eastern Scotland and northern England. Fig. 3.7a shows the early morning infra-red satellite picture with fog penetrating across the lowlands and into the Highland valleys. The model was initialised at 0600 GMT using objectively analysed surface observations, modified

## The use of numerical models in weather forecasting: achievements and prospects

manually in data sparse areas using the satellite picture. By 1500 GMT (Fig. 3.7b) the model cleared the fog from the land except windward coasts allowing temperatures to rise to over 20°C in parts of the Highlands. This created low pressure centres which produced the wind circulation shown and resulted in the clearance of fog from the south side of the Moray Firth. The afternoon satellite picture (Fig. 3.7c) confirms this development. A lesson that has been learnt from this case is the crucial importance of a detailed knowledge of the initial state. Much

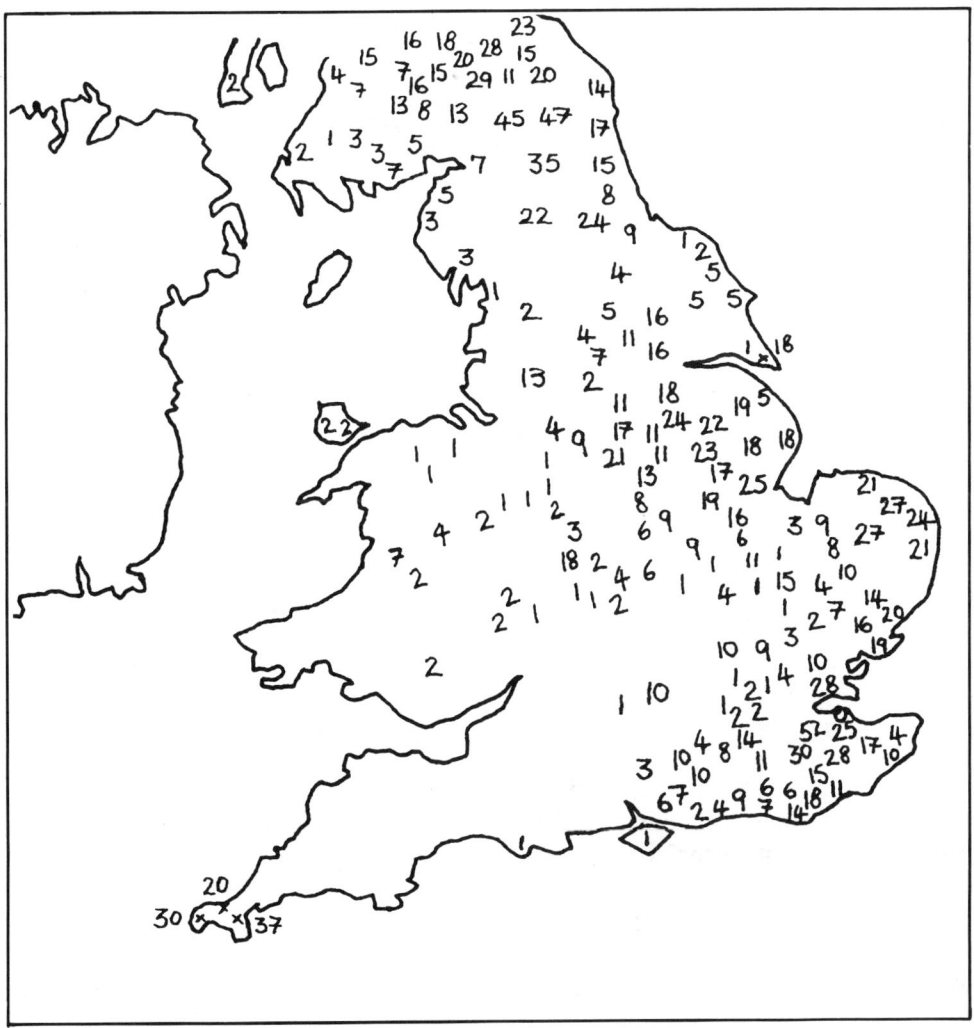

FIG. 3.5. Snow depth distribution at 0900 GMT on 13 January 1987 (a) observed (b) accumulated from successive forecasts.

B. Golding

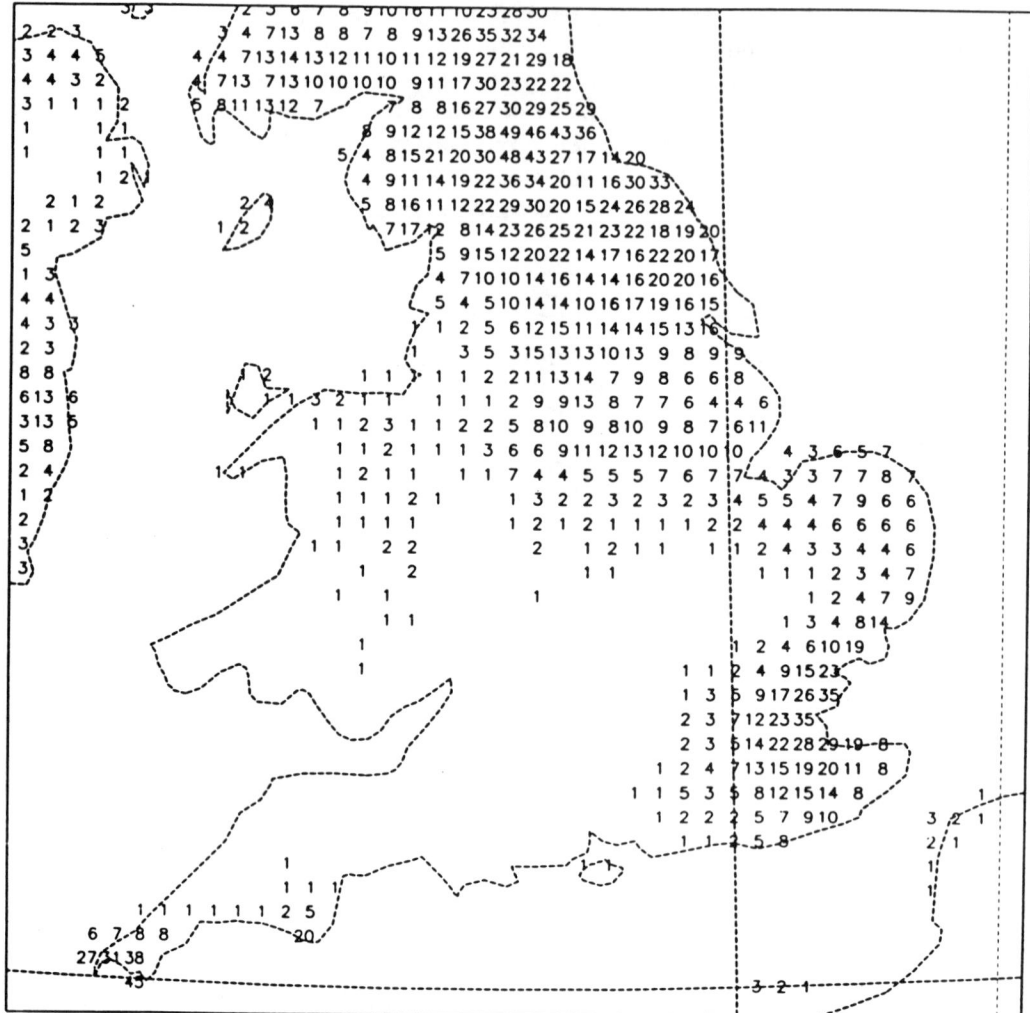

Fig. 3.5. (*continued*)

can be gained from the satellite imagery in this respect but it can only be interpreted correctly in the light of surface and balloon-borne measurements.

## FUTURE PROSPECTS

The science of meteorological modelling never stands still and neither does the computer hardware which supports it. Already, the Cyber 205 supercomputer installed at Bracknell in 1981 has been superseded by faster machines with bigger memories, and in 1988 it is being replaced by an ETA 10 with about 8 times its

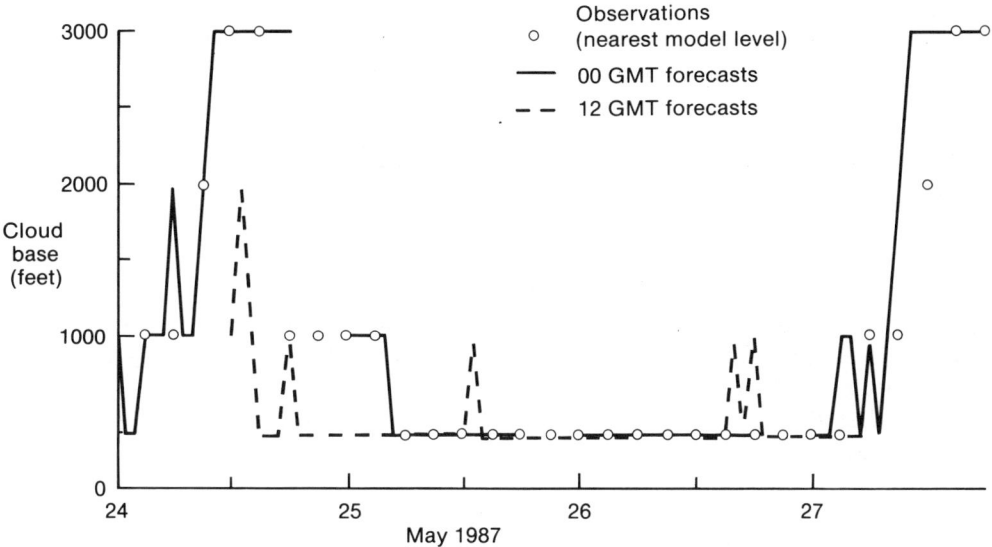

FIG. 3.6. Comparison of observed and model forecast cloud base during the period 24–27 May 1987 at Leuchars. Each forecast lasts 18 hours. Observations have been moved to the nearest model level.

power. Plans for using this computer are well advanced so one can have some confidence in estimating its effect. Firstly, the resolution of the global and regional models will be improved so that smaller scale developments can be taken into account and so that more of the variation of land surface characteristics is represented. Research has already demonstrated the improvement that this will give in the broad scale forecasts, especially at longer range. For the very fine scale model, the forecast domain will be enlarged to include modification of air masses over France and the sea areas around the British Isles, and the vertical resolution will be enhanced to tackle some of the weaknesses in cloud and fog prediction. There will also be sufficient power to give this model operational status when the results justify it. This upgrade will not, however, permit improved resolution of highland topography.

Current projections to the middle of next decade suggest that computers with a thousand times the speed of the Cyber 205 will be available. With that power, it should be possible to resolve at least the lower layers of the atmosphere with a resolution of one or two kilometres, sufficient to represent most hills and valleys. Provided that the inevitable difficulties in representing the physical processes operating at this scale can be overcome, it should provide direct prediction of most of the variables required by users. Given the detail of the surface topography and current experience, wind predictions should be very reliable except, perhaps, with very light flows at night. Given a good representation of cloud, the temperature predictions should also be very good. However, since cloud forecasting will inevitably remain difficult, there will be scope for the human forecaster to modify the guidance so as to minimise errors. This will be particularly important where the customer has a particular sensitivity, for example to frost. The cloud predictions can certainly be expected to improve with better vertical resolution but improved observation of the current state may also be necessary.

As with temperature, the absolute accuracy of the model forecasts may not be the best measure of whether they satisfy the demands of low flying aircraft, for instance. Here there will certainly remain a role for the forecaster judging the likely accuracy of the model against the needs of the user. Similar considerations will probably apply with even greater force to visibility predictions. In some ways assessment of the progress of rain and snow prediction is the most difficult. This is partly because users' requirements differ so much. Even the distinction between rain and no rain has a certain arbitrariness depending on some threshold amount. There is also evidence that this may depend more sensitively on the initial conditions for the forecast, and hence the observations, than some other parameters. Use of satellite

FIG. 3.7. (a) NOAA-8 visible image at 0836 GMT on 27 April 1984. (Supplied courtesy of the University of Dundee). (b) 9-hour model forecast for 1500 GMT 27/4/1984. (c) NOAA-9 visible image at 1448 GMT 27/4/1984. (Supplied courtesy of the University of Dundee).

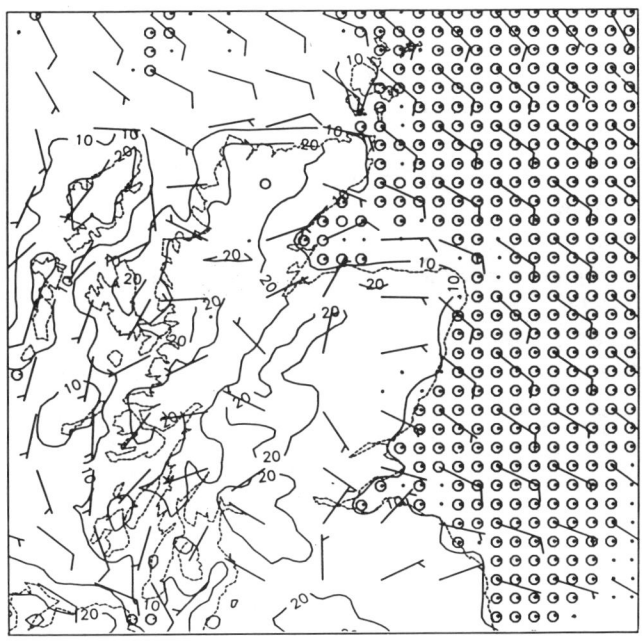

(b)  ○ Fog  ⌒ Surface isotherms
    • Low cloud  ⊢ Surface wind

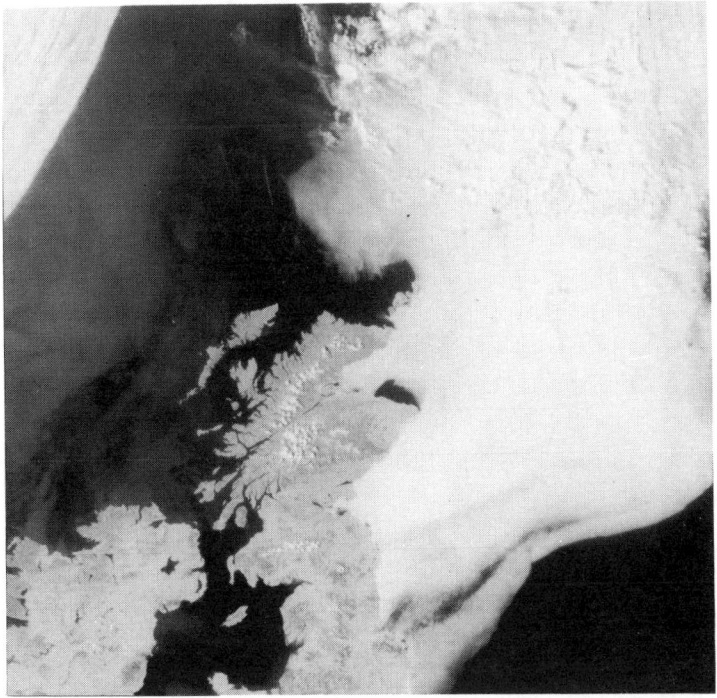

(c)  FIG. 3.7. (*continued*)

data will certainly help in this area. Showers are a particular problem, often being triggered by apparently minor topographic features and yet, on occasion, able to sustain themselves over long distances and times. Some experimental forecasts have produced remarkably accurate predictions of such features suggesting that cautious optimism is appropriate.

In conclusion, the next decade will see further improvements in model forecasts for a day or more ahead, especially in the longer range around a week. In the short range forecasts, model predictions will become the main source of guidance to the forecaster and for some purposes will be used directly. The forecasting of low cloud, visibility and local precipitation will be major challenges in which we can expect some success.

# *In-situ* meteorological monitoring: the practical problem

R. E. W. PETTIFER
*Vaisala (UK) Ltd*

## INTRODUCTION

It has been pointed out that, of all scientific measurements, those of operational meteorology are among the most difficult to make (Pettifer, 1981). There are many reasons for this. Some relate to the instruments which are used for the measurements and some to the ways in which they are used. In many cases, these difficulties are far from obvious and in some instances they can lead the unwary into making bad measurements. The poor quality of these data will often be transparent to the users until they find that decisions based upon the observations they have taken turn out to be wrong. This discussion of the practical problems of making *in-situ* meteorological measurements will consider some of these aspects of the subject and point users at steps which may be taken to avoid them. *In-situ* measurements are here defined as those made by instruments at the point in the atmosphere where they are located. Within this context discussion will be confined to the measurement of meteorological variables on the Earth's surface.

## THE BASIC PROBLEM

There are three characteristics of measurements which, in the particular case of meteorological variables, can present obscure but often severe obstacles to the successful acquisition of good data. These characteristics may be defined in several different ways but, for the purposes of this paper, they are:

>Data Integrity
>Data Representivity and
>Data Reliability

### (a) *Data Integrity*

"Data Integrity" is the extent to which a measurement of a particular variable would agree with a perfect, transfer-standard-quality measurement of the same variable made at the same place and time if it were possible to make such a measurement. There is nothing particularly unusual about the need for this integrity in a measurement. For most measurements it reduces to a simple determination of the extent to which a measurement device is in agreement with some standard, carefully maintained at a Standards Laboratory, and the extent to which a well known code of procedure for the use of the device is observed when it is used.

However, for meteorological measurements, by and large, of these two, the first is not consistently possible and the second is, when available, not generally well known.

Whereas the appropriate fundamental standards against which to judge instruments certainly exist for, say, temperature measurements, they do not exist at all for such important meteorological variables as relative humidity, wind speed, rainfall amount, rate and duration, visibility, snow fall and several more. Moreover, the conventions of measurement for all meteorological variables have been arrived at by long and careful experimentation and, for the most part, are agreed upon (WMO, 1983). However, they are often not self-evident, and in many cases are, therefore, hard to divine for those not trained in the art.

The importance of these problems lies in the fact that a particular meteorological measurement cannot usually be made more than once. This is because that which is being measured is often varying on time scales similar to the time it takes to repeat a measurement, and once a measurement is made, there is, in general, no independent way of establishing its truth. Even the intuitive comparison born of long practice and performed by eye is not available because it is not normally possible to be there when the measurement is made. Moreover, any such assessments would be highly subjective. Consider, for example, the intuitive assessment of air temperature compared to that of, say, a short distance.

(b) *Data Representivity*

"Data Representivity" describes the extent to which measurements made at a particular point are respresentative of the condition of the measured variable over some acceptable scale of space. What is acceptable in this regard depends upon the use to which the data are to be put and there have been many studies published in the technical literature of the subject which address this problem for a wide variety of cases (WMO, 1983).

Once again, however, the field is studded with traps for the unwary. It is to be expected that the measurements of wind speed made near a tall building would in some way be affected by the building. But how far away from the building its influence spreads in this regard and, for a given wind vector, whether that influence is isotropic in space, are questions which do not have self evident answers. Without answers to this type of question, however, the representivity of a piece of data becomes ill-determined and its value depreciates proportionately. From the inspection of a given piece of data, it is almost always impossible to tell what is its quality in this regard. Only by comparing it on many different occasions with other, contemporary data taken over a wide area can this assessment be made (Ponting, 1984). Very few users except the professional meteorological services have the facilities to undertake this type of analysis.

(c) *Data Reliability*

The term "Data Reliability" describes the extent to which a user, probably remote from the measurement site, is confident that the data are correct and to be relied upon. This is a critical feature of *in-situ* meteorological measurements. If they are to justify the expense of obtaining them, the user must be able to use them to make operational, risk decisions. He must, therefore, have complete faith in the measurements. The more unusual the value of the measurement, the greater its departure from some known norm or expected magnitude, the greater is its usefulness provided that the user has the confidence to trust it and is justified in that confidence (Pettifer, 1984).

The task of providing meteorological data which successfully meets the users' needs in respect of these three characteristics falls to the meteorological instrument engineer. And it is not sufficient that he makes a good, reliable device in engineering terms alone. He must understand the way the atmosphere works and how it can be measured.

## MAKING THE MEASUREMENTS

(a) *Instrumentation*

These issues may be illustrated by reference to some features of common meteorological instruments. What is frequently not pointed out to the would-be user, is that in almost every case, an engineering advantage gained in one respect produces a measurement penalty in some other. For example, in the case of a rotating cup anemometer, a strongly-made, massive device which reassures the civil engineer by its robust appearance and general unhandiness, will usually suffer from two serious defects of performance; a relatively high starting speed, which may yield gross inaccuracies at low wind speeds, and a large time constant which will limit its response. On the other hand, small, lightweight anemometers will suffer to a lesser or greater degree from overspeeding which will result in the over-estimate of the mean wind speed value of a gusty wind (HMSO, 1981). These behavioural characteristics of the rotating cup anemometer are enshrined in such obscure characteristics as the distance constant of the device and in benchmarks such as the starting speed but there is no method derived from fundamental principles which can be used to calibrate any rotating cup anemometer. All such devices are "calibrated" by either an empirical measurement in a wind tunnel of some kind, or by assuming that each device manufactured to a given design will behave as did an earlier sample selection of such devices when they were empirically tested in a wind tunnel. The extent to which any particular anemometer design is reliable in this respect is a matter of experienced meteorological judgement or long climatological record obtained with a large population of these anemometers.

The simple raingauge is an important device which has a central place in the field of *in-situ* environmental monitoring. The range of users is wide and ranges from the keen amateur to professional meteorologists and hydrologists. Some of these users are likely to be aware of the serious measurement errors which can be induced by such an obvious influence as wind blowing across a poorly designed or poorly exposed raingauge but, the extent of evaporation losses from these devices, studied by Sevruk (1984), is unlikely to be well known. The probability that, unadvised by some expert authority, the average user will obtain good quality data from a simple raingauge is very small.

*(b) Instrument exposure*

There are some obvious points to be made about the exposure of meteorological instruments. The first is perhaps that what constitutes good exposure for an instrument depends upon the use to which the data from it are to be put. To study the wind regime in the lee of tall trees, it is clearly of no value to put an anemometer in the middle of an otherwise empty field. On the other hand, for a measurement made at a point to be representative of the conditions over a comparatively large area, care must be taken to ensure that the measurements are not influenced by the presence of some local perturbation such as trees, or buildings. Of course, it is almost always necessary to compromise between the desirable and the possible in the siting of instruments. The extent of that compromise and its effect on the quality of the measurements requires judgement which, to be successful, must be based upon considerable experience and skill. It is usually given scant attention by users who do not have meteorological training and whose priorities often lie elsewhere.

As the use of sophisticated meteorological instruments increases, for example to monitor conditions such as poor visibility on roads, the scope for error increases. It would be all too easy for this problem to result in the general discredit of meteorological measurements as operational aids in what have hitherto been non-professional meteorological applications. That would be a pity because, with proper attention to the choice and location of the instruments, these data can make a vital contribution to the optimal solution of operational problems the management of which involves high costs.

## DATA PROCESSING

The variable nature of the atmosphere leads to a further problem associated with the measurement process. It is illustrated in Fig. 4.1., from Lee (1981), in which curve (a) shows the short term variability of the wind vector. Curves (b) and (c) represent attempts to smooth the data over time. With modern electronic devices, a single observation of this variable may occupy only a fraction of a second. It is clear that such a sample is unlikely to be a reliable measure of value integrated over some period of time which is meaningful in the context of, for example, the rate of change of temperature at a road surface. Equally, the simple averaging of a series of such samples is unlikely to provide a fully representative value. It is not even obvious how frequently such samples need to be taken nor how they should be averaged to provide a useful observation. The advent of the microprocessor has provided the instrument engineer with the power to apply almost any sampling and averaging scheme to the measurement problem he faces. But he will not choose correctly unless he knows and understands the physics of the medium he is measuring. The

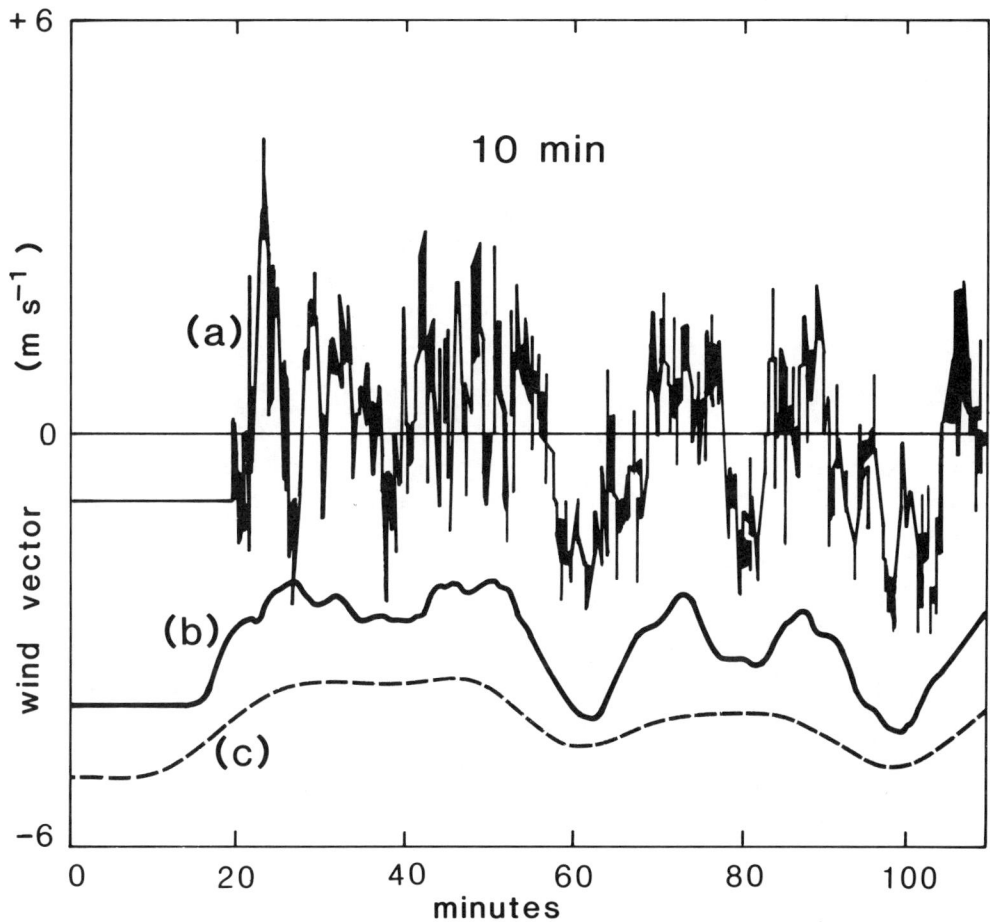

FIG. 4.1. Short-term variability of the wind vector.

result of a poor choice will be bad measurements but often the user, presented with a single output value and no independent means to validate it, will only discover that the data were bad when it is found that the application of the data failed to solve the problem.

Finally, there are many measurements in meteorology which cannot yet be properly treated by *in-situ* instruments. For example, snowfall, water equivalent of snow, cloud amount and cloud type. The lack of accepted techniques for these *in-situ* measurements does nothing to diminish the number of glossy brochures offering instrumental panaceas for these difficult problems.

*Data recovery*

These days the fundamental problems of *in-situ* data logging have been largely solved and good, reliable data logging devices are not hard to find. They are fine if real-time data are not required. But in many cases the value of the data lie in their immediate availability. They rank with daily newspapers as among the most perishable of commodities. In principle, the real-time recovery of data from remote sites is nowadays straightforward. There are available the combined resources of the telephone networks, a geostationary satellite network and a polar orbiting satellite network. However, the scope for data loss or corruption along these high technology pathways seems almost unlimited and great attention to detail, and the nurture of good human relationships, is needed to ensure success. Nevertheless, it must be said that in the majority of cases it is now technically a well understood and straight-forward matter to recover data from remote sites. If we can make good *in-situ* measurements, we can recover them in near real-time.

*Equipment maintenance*

If networks of *in-situ* observing instruments are to be fully operational and cost-effective, it is essential that they are maintained regularly and repaired promptly and efficiently. Attention also needs to be given to the need to undertake on-site calibration, even if only in the form of comparative check values obtained from a reliable alternative system brought to the site occasionally for the purpose.

To support these requirements there are three necessities: a maintenance capability, staffed by competent, well trained personnel, must be available; they must have access to a sufficiently large pool of appropriate spares which have been properly tested and for which there is a properly filled and sensibly short repair pipeline to a competent repair authority; and any on-site calibration or checking must be undertaken by meteorologically skilled observers. This last point has a wider significance than is apparent from the performance of the instruments alone. The observer/engineer must be able to assess the effects of changes in the site on the meteorological quality of the data being produced. Tree growth, the erection of new structures, the removal of existing structures and many other changes can have serious effects upon the representivity and reliability of the data. Equipment and site maintenance are neglected only at the eventual cost of lower data quality.

These support requirements can be an expensive adjunct to the capital cost of installing remote site observing systems, especially automatic systems. Nevertheless, there is no surer way of wasting an investment in such equipment, by depriving the user of access to or confidence in data derived from it, than to neglect these elements of the overall package. It is, therefore, futile to expend resources on any observing network unless sufficient support is also provided to maintain it. It is generally more important to spend resources upon the support infrastructure needed to maintain a small observing network than to enlarge the network without such support.

## CONCLUSIONS

From the above discussion it can be concluded that it is easy to get data from *in-situ* atmospheric monitoring but that it is difficult to get good data reliably. To achieve this objective it is not sufficient merely to choose instruments, even if they are good instruments, fit for the purpose. If the necessary expertise is not available to the user, it is vital that he obtain expert advice on how to deploy, use and maintain the instruments to obtain the measurements he needs. And this advice needs to be a continuing feature of the use of the instruments — calibration, performance and exposures change with time and the gentle degradation of data quality must be guarded against. It should also be concluded that, if meteorological measurements are required, it is worth the effort to get good ones. It can be difficult to do but, with the right support and advice, it can be done.

# SECTION TWO

# TRANSPORT

# Meteorological Office services to transport

### J. G. ALLARDICE

*Meteorological Office, Glasgow Weather Centre*

## INTRODUCTION

THERE are few activities in everyday life that are more weather sensitive than road, rail, air and sea transport. The Met Office has been deeply involved with providing services to the Royal and Merchant Navies since its inception and, since the 1920s, aviation has been the largest commitment. Services to road and rail transport are a product of post-war activity, and for many years have been standardised, low output, warning services. These were never very satisfactory but were based on the customer's requirements and his perception of what could be supplied. The dramatic increase in the accuracy of both short-term and extended range forecasts over the past fifteen years went largely unnoticed by those involved in surface transport, and their perception of the capabilities of the Met Office changed little from that of the immediate post-war years. The continuing advances in the technology and communication systems have transformed the information available to the user within Scotland.

The increased use of surface thermal imaging and the installation of ice detection systems, combined with vastly improved short and medium range weather forecasts, is proving of great benefit to the authorities with responsibilities for road operations. The availability of comprehensive, numerically produced low-level wind forecasts has resulted in a totally revised weather service for helicopters to the North Sea offshore installations from bases within Scotland. Rail operations too have benefitted from the increase in the accuracy of temperature forecasts using model output statistics, combined with standard forecasting practices. Weather information for marine operations are, in the main, not localised, tending to be either trans-Atlantic or even trans-Pacific. There is a well established Oceanic Routing Service available from the Met Office Headquarters Central Forecasting Office. Within a Scottish context, routine inter-Island services use Met Office products, as do many involved in commercial fishing.

It is the intention of this paper to state briefly the position with regard to forecast services available to land and air transport industries in Scotland. There are, of course, services direct to the public such as WEATHERCALL, MARINECALL, WEATHERNET and a vast amount of information disseminated through the media. The paper, however, concentrates on services provided to larger organisations such as local and regional authorities, British Rail and the Civil Aviation Authority and, through them, to the general aviation user.

## ROAD TRANSPORT

Services to the road transport industry are, in the main, concentrated in the six winter months November to April (Fig. 5.1). Although this graph does represent individual enquiries at one Weather Centre it is a guide to the weather sensitivity of road users for whom the regional highway authorities have a responsibility. Until 1984/1985 the weather information supplied to the authorities, to assist them in their

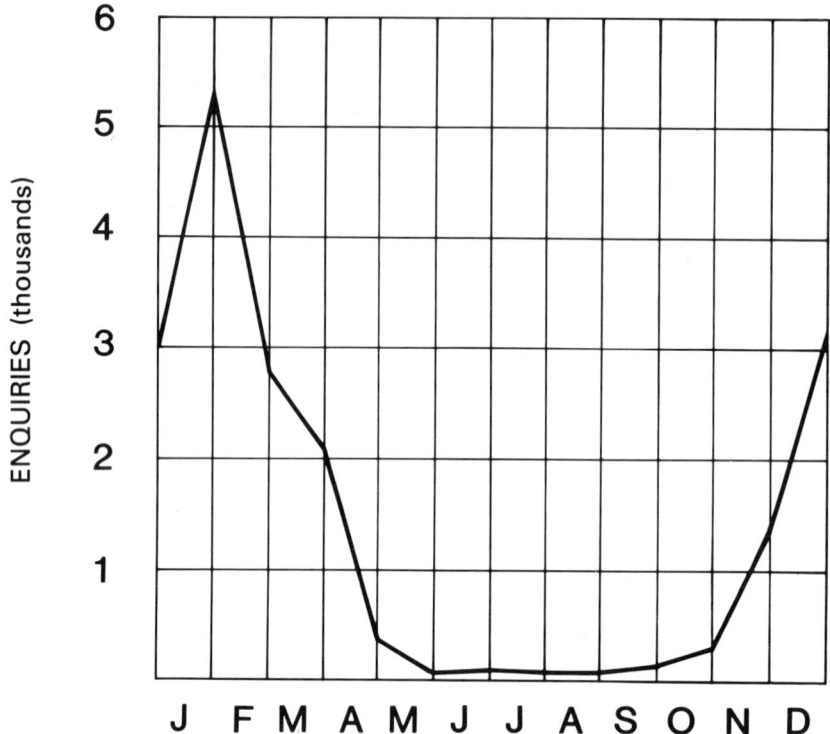

FIG. 5.1. Mean monthly road transport enquiries received at the Glasgow Weather Centre (1978–87).

decision-making with regard to winter road maintenance, tended to be brief and pessimistic. Without exception it consisted of specific element warnings such as the likelihood of ice or snow. There were a number of drawbacks:
1. The warning applied to complete administrative divisions and, if a low probability of adverse conditions existed in one section, a warning was issued. In the event of no adverse conditions, there was no contact.
2. The forecaster was actively seeking the worst scenario, so the warnings tended to be pessimistic.
3. There was virtually no feedback, so the users' dissatisfaction was seldom apparent to the forecaster.
4. Forecasts were valid within the 24-hour period, and in many instances within 18 hours.

This type of service was not held in high regard by the majority of users. It was, however, inexpensive and it was monitored and updated as necessary throughout the 24 hours.

During 1985 the Met Office reviewed this long-standing service and a number of alternative options were tested. In Scotland, with the co-operation of the highways authorities, a totally new approach was adopted. It was always the case that using administrative areas had not been ideal and a move towards more meteorologically significant zones was initiated. The warning services ceased and a daily comprehensive divisional forecast provided, including a forward planning outlook up to 5 days ahead for the whole region. After the first winter the consensus was that it had been moderately successful but required further improvement. Similar services were also trialed with counties in England. Coincidentally, the regional authorities, seeking more cost-effective ways of operating, were in the process of utilising new technology, and a considerable number had thermal imaging surveys carried out over the roadway networks. By doing so, areas with unique temperature characteristics could be identified. Modern ice detection systems linked to easily accessible master stations were being installed and are continuing to be established (Harverson, 1985). As shown in Fig. 5.2, there is now a wide availability of ice detection sites in Scotland and the logical approach was to produce forecast road temperature profiles for specific sensor sites. Under the aegis of the Department of Transport, trials were conducted at Birmingham Airport Meteorological Office of a numerical model of the process developed at Birmingham University. These tests proved successful and the Department of Transport developed a specification for the necessary computer-to-computer communications between sensor systems and the Met Office. This protocol has been updated in the light of experience over the ensuing years since 1985.

Having reviewed the various forecast trials, the Met Office produced a comprehensive service called OPEN ROAD, which included the elements that the county and regional authorities had perceived to be of most use, and this was launched for the winter 1986/1987. The components of the package comprised the following:

A comprehensive 24-hour forecast concentrating on likely adverse conditions (ice,

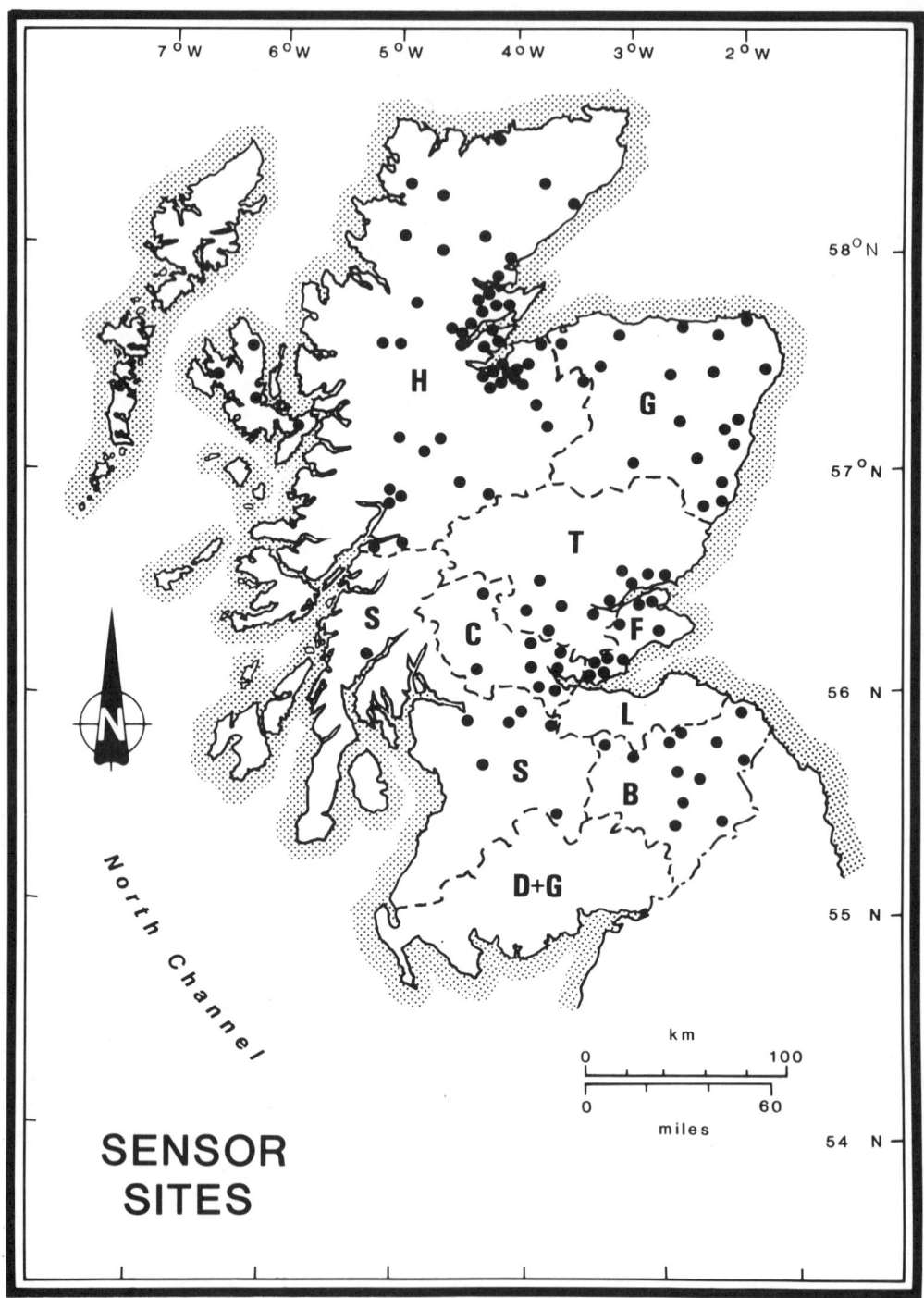

Fig. 5.2. Location of sensors for detection of road ice in Scotland.

snow, freezing rain), including forecast minimum and maximum temperatures for air and road surfaces.

An adverse weather outlook for the specific area/division for days 2 to 5.

Forecast road surface temperature profiles for one or more sites.

A 24 hour consultancy service allowing direct contact to a forecaster.

24 hour monitoring with set update times at the Weather Centre to allow required amendments.

In England and Wales there is also an option to receive weather radar data from the network. This is transmitted by DOCFAX and carries interpretative comments on likely movement and development of precipitation areas. With the establishment of further weather radars this facility should be available in parts of Scotland within the next two years.

The production and provision of forecast surface road temperature profiles followed the establishment of road sensors, and a further development from thermal imaging surveys has been the facility to receive forecast thermal maps. This too is available to certain county authorities in England as part of an enhanced OPEN ROAD service. The service to county and regional highway authorities is undergoing continual development and it is acknowledged that much of the success so far stems from the close co-operation between the Met Office, the Scottish Development Department and the regional highway authorities. The expertise of those companies involved in thermal imaging and road sensor development has been invaluable.

## RAIL TRANSPORT

Though Scotland, in general, suffers more extreme weather conditions than other parts of the British Isles, the railway system is probably the mode of surface transport most tolerant of adverse weather. The provision of weather information for British Rail, like that for the roads, had changed little over the last twenty to thirty years. During the winter this too has tended to be based on pessimistic warning services for specific weather elements, such as rail ice, frost, snow and fog and, in the case of overhead power cable operation, an interest in strong winds. Due to the rather generalised approach, where any risk of adverse conditions within an area warranted a warning issue, the service was perceived to be unreliable and held in low regard by the operators.

With the development of the revised services for highway authorities in 1986, there was obviously a case for a similar type of service directed towards the railways. Coincidentally, during the same year the British Rail Board and the Cranfield Institute of Technology had published reports attributing losses of revenue due to unpunctuality in the £1 million per annum range. Adverse weather was identified as one cause. There was therefore motivation on both sides to revise procedures, and subsequent discussions between Scientific Services BR (Scotland) and the Met Office resulted in a totally revised service being tested in Scotland during January to March 1987. This new service was also based on what was determined to be more meteorologically significant areas and an attempt was made to forecast specifically for the lines involved, rather than for the generalised areas that had previously been the case. One problem entailed identifying those parts of the rail network likely to be subject to specific weather-related delays. Tracks running along shore lines were a consideration, where inundation by waves or over-exposure to strong winds could be an identifiable problem. It was not feasible to distinguish between individual cuttings and raised lines but high and low level trackings could be identified, as could sheltered lines. The forecasts contain similar elements to OPEN ROAD but are much briefer and in the case of ground temperature, refer to a grass temperature rather than a road surface. The delivery method is by TELEX to British Rail's Central Communication System for re-distribution. Further work is currently being carried out within the Met Office on modelling rail temperatures.

It is interesting to note that, although the road and rail services have similarities and the aim in both cases is to maintain a free flow of traffic, the direct use of the data is not the same. In the case of the highways it is primarily used for ensuring that salt and grit are spread in a timely and economic manner and that personnel and plant are correctly placed for road clearance operations. Although the railways do use the information for line maintenance, a large part of the effort is directed towards the plant and engineering side in ensuring the rail stock is in operating condition in sub-zero temperatures.

The cessation of the winter services in April does not end the supply of information to ScotRail. During the ensuing period up until autumn a daily service is provided to British Rail (Scotland) Control. In hot weather it is relatively easy for the rail in direct sunlight to reach comparatively high temperatures, especially in cuttings, with shelter from any cooling effect of wind. With rail temperatures likely to exceed 32°C, work on the rail bed is curtailed and speed restrictions may also be applied due to the possibility of distortion caused by unequal expansion and contraction. Forecasts of wind, air temperature, cloud cover and likely precipitation are supplied for all relevant areas for days 1 to 3 ahead to British Rail Control. This service runs from spring through to late autumn.

## AIR TRANSPORT

The supply of weather information to general aviation and commercial operators in Scotland has undergone significant redevelopment during 1987. The introduction of AIRMET for the private and smaller commercial operator has totally changed the method of delivery and structure of aviation forecasts. For those operating at flight levels below 15,000 ft (4570 m) AIRMET replaces both route and local areas forecasts. Scotland is subdivided into a number of areas which are, as far as is possible, meteorologically distinct (Fig. 5.3). The forecasts are issued four times a day and are valid for an eight hour period. The format is standard, including:

a general description of the synoptic situation,

the likely significant weather both spatially and with time,

*Meteorological Office services to transport*

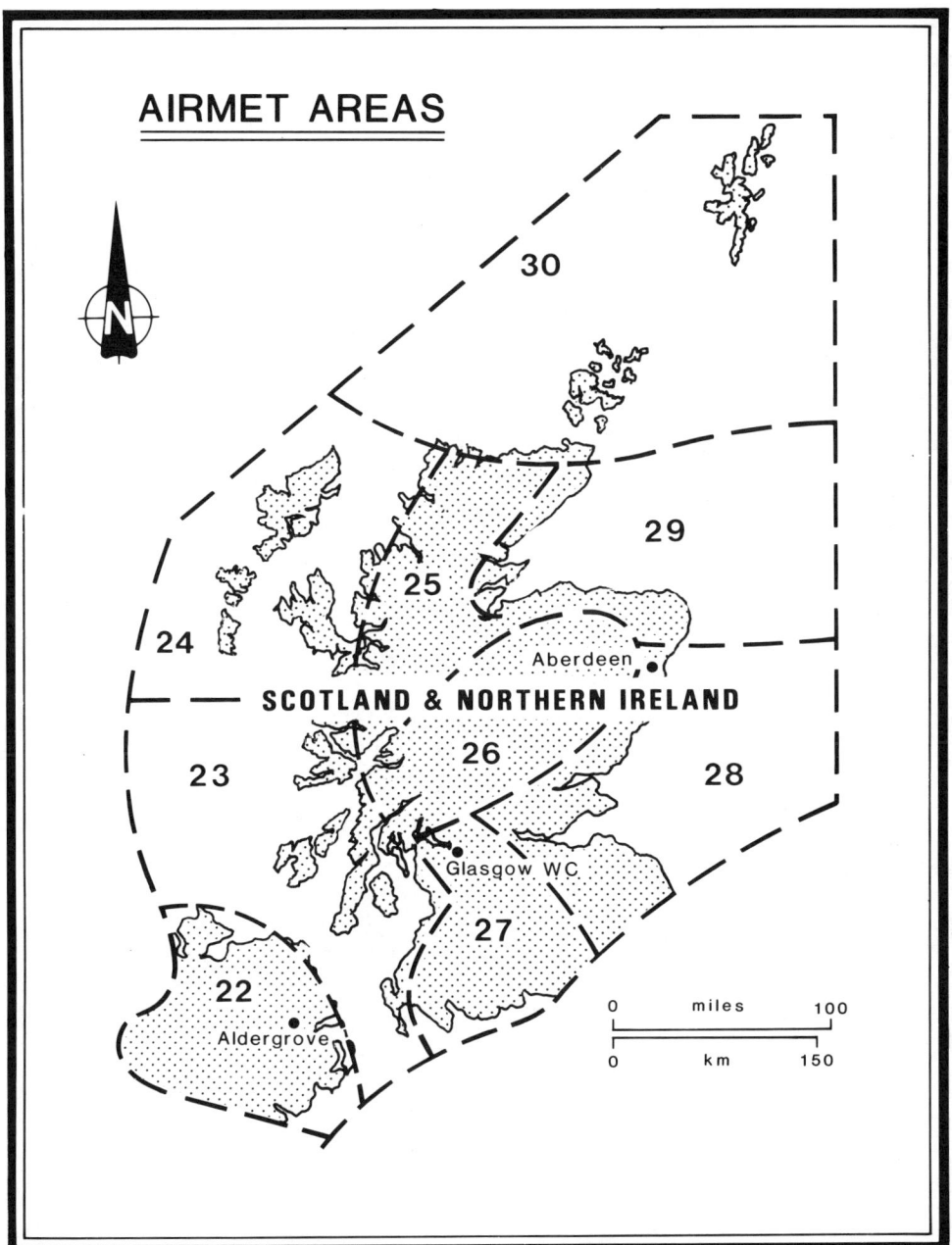

Fig. 5.3. Division of Scotland into AIRMET forecast areas.

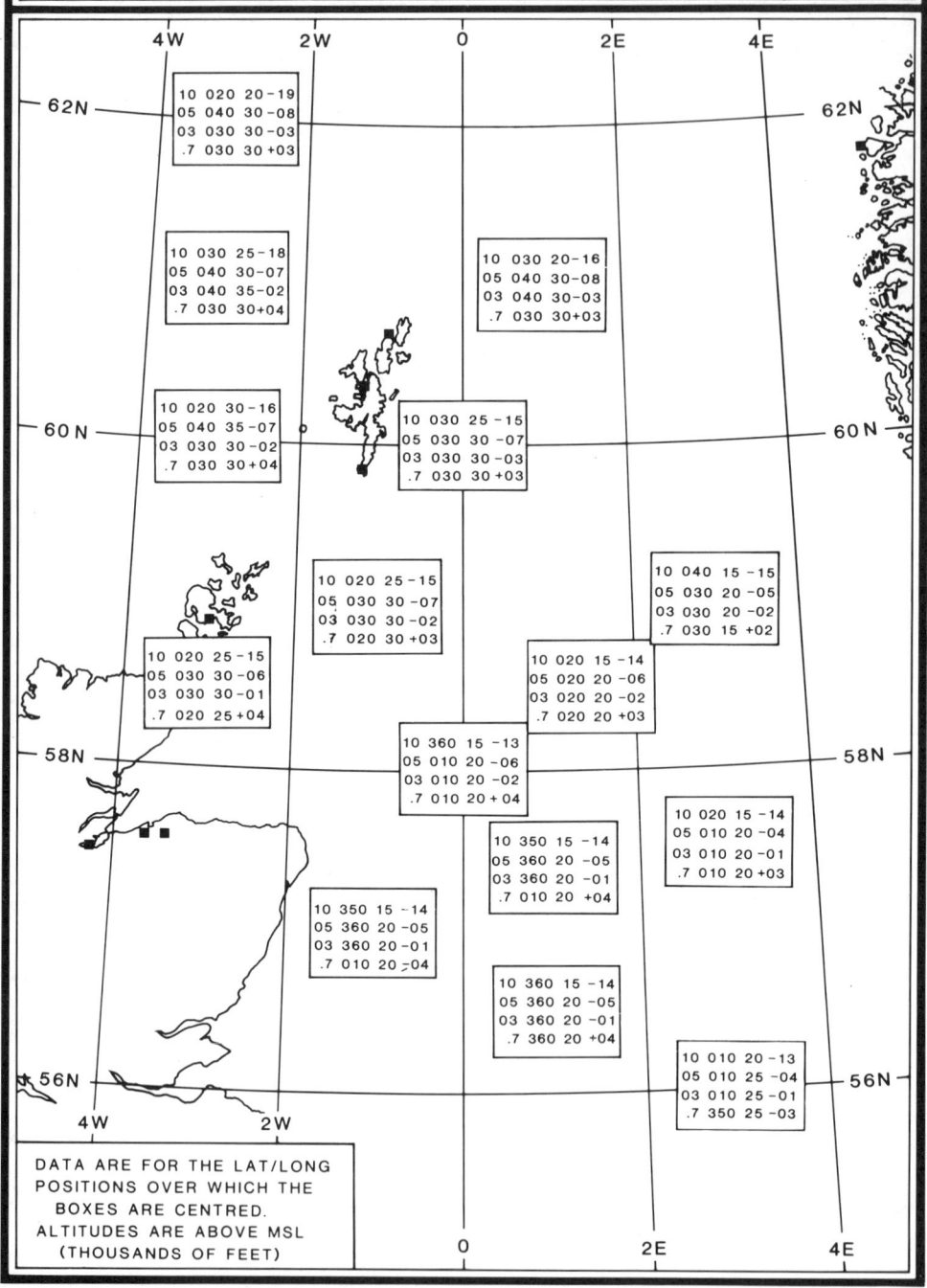

standard elements as required by civil aviation convention including inflight warnings of specified adverse conditions, and

forecast upper winds and temperatures at fixed flight levels from surface to 18,000 ft (5482 m) together with cloud conditions and visibilities in each area. The information is available to users through hard copy documentation at flight briefing units on major airfields, by TELEX or by the fixed telecommunications network for aviation. The forecasts are also available through a premium rate telephone service, with full text read at dictation speed.

For the more specialised user, especially the helicopter operators to offshore installations in the northern North Sea, recent developments have been the reverse of AIRMET. Here the recent innovation has been the move away from strictly text based forecasts to a mostly graphical output. Since the great majority of movements take place from Aberdeen, the main destinations are identified by a number of radials. Numerically produced forecast upper winds and temperatures are given for specific points on the individual radials (Fig. 5.4). The winds at these positions are interpolated from grid points on the Meteorological Office fine mesh model and the data is produced in 3-hourly steps for four set flight levels between 700 ft (213 m) and 10,000 ft (3045 m). A manually produced significant chart for a mid point time in the forecast period supplements the numerically produced data and these, together with a brief text, form a very comprehensive brief. Multi-address document faximile is the communication method, and subsequent radio linked faximile transmissions to offshore installations have proved highly successful.

Currently, the two morning forecast issues are based on 0001Z data while those in the late afternoon and evening use the 1200Z data based output. In the event of a communications or hardware failure, each run produces sufficient output to enable a further two issues of spot wind forecasts. The detail afforded by this method allows 60 grid point forecast winds for each fixed time giving upwards of 250 winds with each issue. Without the automated spot wind issues, the service would not be feasible.

One drawback that was envisaged was a requirement to manually amend should the numerically produced data prove suspect. This would be virtually impossible given the amount of information, but an alternative of reversion to a text output was established. It has not proved necessary to rely on this fall-back position. During the four winter months since the service was established, the forecast winds have proved extremely reliable with only the lowest 700 ft (213 m) wind needing manual adjustment on occasions. It is foreseen that, on some occasion, the predicted pattern may undergo a major disruption should a small rapidly moving feature move through the field but this too could be handled by reverting to the issue of

---

FIG. 5.4. An example of the forecast output for upper winds and temperatures available for helicopter operations over the North Sea. From left to right the columns in the located boxes indicate fixed forecast heights (700, 3000, 5000 and 10,000 feet), windspeeds, wind directions and temperatures respectively.

individual route forecasts through the area likely to be affected for the short time period this is likely to be required.

Both AIRMET and the revised helicopter forecasts are in their infancy but they have replaced outdated services which have been in operation for about 15 years or more. It is hoped that these new services for aviation will also be accepted for a considerable period before they too are replaced due to changing requirements and advances in technology. The increasing use of automated documentation would suggest that the supply of weather information to commercial and private aviation will, within a fairly short time span, be almost totally in a graphics form.

## OTHER SERVICES

In addition to these standard services to land transport and aviation there are numerous individually tailored services for organisations representing road users. The Automobile Association and Royal Automobile Club receive both national and local weather information on a daily basis. The BBC Motoring Unit has a longstanding arrangement whereby warnings of specific elements, such as strong cross winds, heavy rain or ice and snow, are issued. During a winter season the variety of weather in Scotland virtually demands that both Aberdeen and Glasgow Met Offices are involved in a number of issues to the BBC daily. Individual commercial road transport firms and retail organisations also make use of weather information for both daily operation and forward delivery planning. The various police forces throughout Scotland have a critical interest in adverse weather and they too have warning arrangements with their local Met Office.

At the present time, services to and enquiries from land, sea and air transport probably amount to about 35% of the workload at Weather Centres involved in catering to public and civil aviation requirements. The demand for the information, the expertise involved and the capital outlay has increased steadily over the past five years and is likely to continue this trend for the immediate future.

ACKNOWLEDGEMENT. The author wishes to thank Miss D. Harverson, Findlay Irvine Ltd, for assistance with ice sensor site data.

# The response of a Regional Roads Department to adverse weather

D. L. BRINHAM

*Department of Roads and Transportation, Central Regional Council*

## INTRODUCTION

CENTRAL Region is strategically placed between the Clyde and Forth. Consequently, as well as coping with the traffic generated within its own borders, it provides access from north and south for a large volume of through traffic, both private and commercial. It is approximately 2622 km$^2$ in area with a population of 272,000. The bulk of the population is centred around the Falkirk/Grangemouth industrial complex and Stirling, with the balance spread over the rest of the region (Fig. 6.1.). Geographically, the Region's roads network ranges from close to sea-level in the east to road levels above 300 m in the north-west.

The Region has three Districts, Falkirk, Stirling and Clackmannan. Falkirk, lying to the south of the Forth Estuary, is densely populated in parts with substantial industrial content. Easterly winds can bring severe conditions to this area and varied road maintenance techniques are required to deal with these. The petrochemical complex at Grangemouth can raise surrounding air temperatures by approximately 1·5°C and this has a beneficial effect on road surface conditions. Stirling District forms the largest district in area with the town of Stirling at its eastern end. To the west and north the region is predominantly rural with the road network serving an area of hills and lochs which are vulnerable to inclement weather. Highly localised places can have marked variations in winter weather conditions which depend on factors such as altitude and geographical location, especially in relation to nearby hills. The winds are also influenced by local topographic features. Experience has shown, for example, a marked increase in weather severity between Stirling and Bridge of Allan, Bridge of Allan and Dunblane, and Dunblane and Balhaldie along the A9.

## RESPONSIBILITIES

There are a number of conditions which require a response in addition to the obvious problems of frost and snow. *Wind* can disturb fallen leaves and block drainage offlets with resultant flooding, or it can blow down trees, and damage signs and lights, causing disruption to road users. *Rain* can cause flooding, land erosion or landslips and consequent breaking up of the road surface, or more simply the insidious corrosion of steelwork on bridges. *Frost,* apart from the obvious travel hazards and inconvenience, can, if prolonged and severe, result in millions of pounds worth of structural damage to roads and bridges. *Long dry spells* may cause the

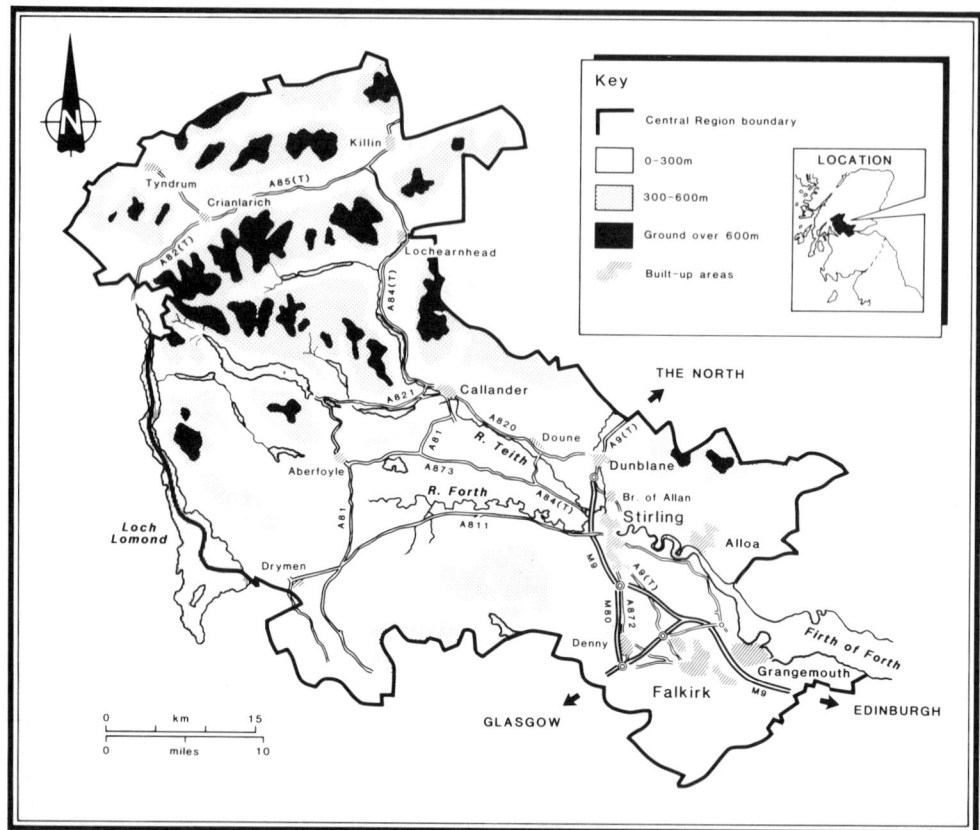

FIG. 6.1. Topography, settlement and the major road network in Central Region.

bitumen on surface dressed sections of road to start to bleed, resulting in skidding conditions. The sub-grade can dry out which results in settlement and cracking of road surfaces. In the planning of road maintenance works, such as resurfacing and surface dressing, adverse hot weather conditions have to be catered for. However, this paper will concentrate on winter snow and icing conditions since this is a problem throughout Scotland and accounts for, on average, 17 per cent of the annual road maintenance budget.

Central Regional Council is, as far is as practical, legally responsible for maintaining roads within the region. The Roads (Scotland) Act 1984 provides the guidelines for all matters relating to roads and footways... "A local Roads Authority shall manage and maintain all such roads in their area as are for the time being entered in a List of Public Roads prepared and kept by them under this section; for the purposes of such management and maintenance they shall, subject to the provisions of the Act, have power to reconstruct, alter, widen, improve or renew any such road or to determine the means by which the public right of passage over it, or over any part of it, may be exercised."

There are different categories of roads throughout the Region, as there are throughout Scotland. Central Regional Council is responsible for 2045 km of roads. Of this mileage 1529 km (75%) are totally financially maintained by the Regional Council. That is, the cost of maintaining these roads is met from the Rates Support Grant. These roads vary in category from principal class routes to classes 2 and 3, and unclassified roads. Their importance can vary from sections of busy dual carriageway to a small cul-de-sac in a housing estate. Although these routes are categorised on the basis of importance by traffic volumes and locations, the householder living in the cul-de-sac would probably disagree with the official assessment when struggling to get his car out onto the distributor road during a snowstorm.

The trunk roads basically form the spines of the regional roads system and incorporate the motorways, sections of dual carriageways without motorway classification, and single carriageways which are nevertheless trunk roads, such as the A84/85 from Stirling through Callander, Crianlarich and eventually to the regional boundary. Lengthy sections of the Region's trunk roads are at considerable heights above sea-level and the Scottish Development Department (SDD), on the basis of information regarding temperatures and snowfall frequencies obtained from relevant weather stations over a period of years, calculate funding for winter maintenance and flooding using a formula applicable to the whole of Scotland. Payments are then made to the Region to cover the costs incurred by the Roads Authority for snow clearing and salting on trunk roads. The trunk roads differ from Regional roads in that the cost of their construction and maintenance is wholly borne by the Secretary of State through the Roads Division of the SDD. The actual maintenance work is carried out by Central Regional Council Roads Department on an agency basis and the necessary finance is provided by the SDD. For winter maintenance, as with other trunk road related functions, the SDD stipulates the amount to be spent and these limits must be adhered to unless the SDD itself approves any variations.

As an aid to winter maintenance, previous experience is now reinforced by data from *thermal mapping* and ice detection sensors which provide up-to-date information on sections of roadway known to be more vulnerable to adverse weather conditions. In addition, snow fencing is erected at sites which are susceptible to drifting snow. This work is carried out in September/October after access to the adjoining fields has been agreed with adjacent proprietors.

## FINANCIAL ASPECTS

As has already been stated, the cost of winter maintenance operations over the past 9 years has represented an average of 17 per cent of total expenditure (Table 6.1). Changes in procedure, including increased use of the Icelert System, however, sees the current year's budget (1987/88) at £850,000, which is only 14 per cent of total

TABLE 6.1. *Central Regional Council Roads & Transportation Department*
Winter Maintenance Statistics

|  | Regional roads | | Trunk roads | | | |
| --- | --- | --- | --- | --- | --- | --- |
| Year | Winter £'000s | Total £'000s | Winter £'000s | Total £'000s | Salt used tonnes | Comment |
| 1980–81 | 429 | 3667 | 154 | 981 | 11,800 | |
| 1981–82 | 822 | 3785 | 222 | 1606 | 20,000 | |
| 1982–83 | 603 | 5911 | 246 | 2327 | 18,200 | Grant Aid |
| 1983–84 | 1239 | 5877 | 427 | 2526 | 33,000 | |
| 1984–85 | 1075 | 5114 | 295 | 2109 | 17,400 | |
| 1985–86 | 1111 | 6105 | 431 | 1874 | 22,000 | |
| 1986–87 | 896 | 6169 | 382 | 1610 | 15,600 | |
| 1987–88 | 850 | 6540 | 235 | 1159 | 17,000 | Estimated |

Current Salt Stock–33,500 tonnes
Length of Trunk and Motorways Maintained–247 km
Length of Regional Roads Maintained–1935 km
Length of Footways Maintained–1710 km
Length of Footpaths Maintained–220 km

expenditure (Table 6.2). This is based on the use of 19 gritters with 9 back-up units for snow clearance and is calculated on 30 days and 10 evenings of frost, and 20 days and 10 evenings of snow, using 700 grit bins and 12,000 tonnes of salt on the Region's roads. In exceptionally bad winters, budget allocations on both Regional and trunk roads are frequently exceeded and supplementary allocations have to be sought. Additional money generally has been made available from both the Regional Council and the SDD. While no direct financial contributions are received from District Councils they make their labour available free of charge in adverse conditions to treat footways and footpaths, salt being supplied to them for this

TABLE 6.2. *Central Regional Council Roads & Transportation Department*
Winter Maintenance Costs 1986/87

| Class of road | Expenditure £ | Carriageway length (km) | Salt used, tonnes | Unit cost/km £ |
| --- | --- | --- | --- | --- |
| Trunk/Motorway | 282,000 | 247 | 5000 | 1546·56 |
| 1 | 292,000 | 357 | 3850 | 817·93 |
| 2 | 165,000 | 285 | 2400 | 578·95 |
| 3 | 73,000 | 332 | 850 | 219·88 |
| Unclassified | 366,000 | 960 | 3500 | 381·25 |

purpose. Ploughs fitted to footway sweepers and tractors are particularly useful for clearing snow from footways.

Rock salt is the main commodity used in winter roads treatment and the Region is recommended to store 25 tonnes per kilometre of carriageway for treatment of trunk and principal roads, and 10 tonnes per kilometre for all other roads, so a total stock of 33,000 tonnes is currently (1988) held. The cost of this is £570,000, the salt being purchased through an annual tender system, mainly from ICI and increasingly from Sicily, Algeria and West Germany.

With a financial outlay of this nature, limiting the wastage of salt is of prime importance. The principal loss of salt during storage is due to rainfall which, apart from leaching out the salt and reducing its strength, also removes the added anti-caking inhibitor. This results in the formation of lumps which, if not removed, choke the gritters. Salt loss on an open pile equates to $0.05\,\text{kg tonne}^{-1}$ mm of rainfall. Taking into account an annual rainfall that, within the Region, varies from 900 mm per year in the south and east to 2400 mm in the north and west (Smith, 1974; Harrison, 1980), losses of 2145 tonnes per year, costing £42,900, could be expected.

The use of reinforced polythene covers was introduced a few years ago to mitigate this loss but their predicted reasonable life expectancy has not materialised and difficulties in using them have been experienced, especially in high winds and heavy snow. Salt barns have now been introduced at two depots and a further two are planned, with more to follow. The use of salt barns also lends itself to the storage of low moisture salt which produces a saving by giving a better rate of spread on the ground. Barns with 4000 to 6000 tonnes capacity cost approximately £90,000 each and have been part financed by SDD. It is estimated that, allowing for the capital interest repayments, rates, increased cost of low moisture salt, and maintenance of the structures, they should pay for themselves within 7 to 10 years.

## MANPOWER AND EQUIPMENT

The Department of Roads and Transportation has a total staffing level of approximately 200, of whom some 35 are directly involved in winter maintenance. In addition, the total labour force is some 365, 280 of whom are available for winter maintenance duties. Within the Region there are 29 routed gritters with a further 11 sitting ready for immediate use. In addition, there are a further 11 which can be mounted in severe conditions. There are also 18 trailed gritters which can be mobilised in the worst conditions. Other equipment available for use includes snow blowers, loadings shovels, JCB's, footpath snowploughs, tractors with plough-blades and footpath gritters. In normal gritting conditions these vehicles are operated single handed except in the north of the region where, due to poor radio reception, they are double-manned.

The snow blowers and four loading shovels are supplied under a winter maintenance contract to supplement the plant held by the Region. The blowers are primarily of use on the higher levels of road where the snow is drier and more powdery. At lower levels the wetter snow tends to clog the exit chute from the blower, making it less functional.

Salt will melt ice at temperatures below 0°C but loses its ability to do so as the temperature drops towards −10°C. To be effective, salt must be spread evenly and at rates to suit prevailing weather conditions. Excessive salting is undesirable on both environmental and economic grounds, and hence all gritters are calibrated, and spread rates are clearly marked. Pre-gritting is normally carried out at a target spread rate of $10\,\mathrm{g\,m^{-2}}$. Where ice has already formed, spread rates are increased to up $40\,\mathrm{g\,m^{-2}}$, depending on the amount of ice and prevailing temperature. The salt treatment required for melting snow is also $40\,\mathrm{g\,m^{-2}}$. Ploughing, however, is recommended as soon as snow depths exceed 30 to 40 mm with each pass supplemented by salt applied at $10\,\mathrm{g\,m^{-2}}$. Should the temperature drop and there is a need for ploughing operations to continue, it is important to monitor temperatures so that spread rates may be increased up to $40\,\mathrm{g\,m^{-2}}$.

As already indicated, Central Regional Council has installed a computerised ice warning system as an aid to improving the standard of its winter maintenance operations while achieving an overall reduction in its budget. The Findlay Irvine Mark 5 Icelert System consists of eight data gathering outstations located at strategic points throughout the Region. These stations utilise sensors to collect information on air temperature, dew point temperature, wind speed and wind direction, together with road surface conditions, including surface temperature, state of surface, and the residual salinity. When the duty officer requires information he can instruct the master station, which includes an IBM micro-computer, in the central office in Stirling to contact any or all of the outstations and display current data. In addition, if any outstation detects a change in conditions outwith specified limits it will automatically inform the master station, which will trigger an audible and visible alarm.

Over and above the master station, each system can support an indefinite number of secondary master stations. These secondaries present the same data as the master station but are in the form of portable micro-computers so that they may be taken home overnight by the duty officer on call. The data displayed on the VDU screen shows sensor location, status (normal, standby or critical), ground temperature, air temperature, surface condition (wet, dry, ice), whether salt is present, wind speed and direction (optional) and dew point temperature (optional). The Met Office currently access data for two of the outstations and using this data are able, with their own predictive models, to give both realistic and pessimistic projected temperature graphs for these locations. These graphs are transmitted from the Met Office direct to the master station in Central Regional Council's offices in Stirling. Weather forecast data is received by Telex during normal working hours, while the secondary master stations are able to access the most up-to-date forecast for the area at any time via British Telecom's service *Telecom Gold*. This allows duty officers to remain fully acquainted with a changing forecast during the night and at weekends. The indications are that substantial savings are being made but it is too early to fully quantify these.

*The response of a regional roads department to adverse weather*

## THE WINTER MAINTENANCE SERVICE

Central Regional Council's Department of Roads and Transportation is responsible for providing the winter maintenance service on adopted highways throughout the Region. This is based on identified mileages of varying priority route categories as indicated in Table 6.3. In order to ensure a consistency of action across the Region and a controlled labour response to operational requirements, a two-tier system of standby supervision is adopted for the winter period. Principal Engineers have specific responsibilities for predetermining required action on the various routes and authorising treatment in accordance with policy constraints. During working hours these two engineers evaluate weather forecasts and road condition data and, in the light of these, determine a suitable course of action for the following overnight period. This decision is communicated to the duty officer who then initiates any necessary action through the duty supervisors.

Commencing at the end of October, and for the required duration of the Regional Council's winter maintenance service, responsibility for the management and control of the service outwith normal hours rests with duty officers and supervisors. The Regional Council operate a 24 hour Emergency Control Centre in Stirling. All calls relating to winter maintenance are passed initially to the duty officer. Having initiated action the duty officer passes operational control to the duty supervisors. They are required to monitor predicted weather forecasts and the status of the plant and vehicles. On the basis of this information they investigate the laid down

TABLE 6.3. *Operational priorities on roads in Central Region, Scotland*

| Priority | Roads | Action |
|---|---|---|
| 1A | Motorways<br>*Slip roads*<br>*Interchanges*<br>Trunk Roads | Precautionary salt treatment when appropriate and clearance of ice and snow 24 hrs per day |
| 1B | Regional Roads<br>*Major bus routes*<br>*Urban traffic routes* | Precautionary salt treatment between 05·30 and 22·00. Snow clearance 24 hrs per day |
| 2 | Secondary Routes<br>*Urban spine roads*<br>*Rural roads serving a reasonable population* | Treatment when the general outlook is for ice and snow beyond midday. |
| 3 | Tertiary Routes<br>*Urban and rural routes of lesser importance* | Treatment only carried out when ice/snow has persisted for more than 48 hrs and the general outlook is for no thaw |
| 4 | Remainder of network | Treatment only in very exceptional weather conditions |

procedure to cope with all aspects of the weather, in close liaison with management. At the end of each shift they record action taken and materials used, and list equipment available for use.

## CONCLUSION

New technology, such as the recent installation of the Icelert detection system has greatly enhanced response to icing conditions and has led to genuine savings by eliminating superfluous salting and unnecessary call outs. The Met Office forecasts are now taken very seriously and, in conjunction with Icelert, form the basis for decisions on action to be taken. Central Regional Council look forward to the introduction of weather radar to Scotland for this, with its proposed visual display systems, should give added effectiveness to the winter treatment of roads.

However, in spite of this new technology the Region value the information available from police patrols driving along the roads affected, and the local roadmen who have developed, without doubt, an unscientific but nevertheless remarkably accurate nose for the weather. Finally, no matter how well the roads have been treated, the question of the driver's safety still rests with his own ability to allow for the conditions and adjust his driving accordingly.

# The effect of severe weather on the Scottish rail system

C. L. CRAWFORD

*ScotRail, Glasgow*

## INTRODUCTION

RAILWAYS have a reputation for reliability in bad weather which usually brings additional traffic from other modes of transport. The basic railway structure presents the opportunity to run a safe and efficient service which, although not immune to the vagaries of bad weather, can quickly recover from adverse situations to allow a virtually normal timetable pattern to be achieved. Some weather conditions, such as fog, affect railway operations very little whilst snow can still have a major impact. This paper examines how adverse weather is and can be detected, the most important types of severe weather, and how the train service is managed on a contingency basis.

The ScotRail system has over 1500 route miles and currently employs nearly 14,000 staff (Table 7.1). Rail movements include both short and long distances. Over the ten years 1977 to 1987 there has been a considerable reduction in staffing from 19,189 to current levels. The creation of modern Signalling Centres covering wide areas of the country has led to the elimination of 156 signal boxes. Other staff reductions have resulted from driver-only operation on selected freight and passenger routes, and the substantial destaffing of stations, together with generally increased mechanisation and restructuring of local administration. These reductions have considerable repercussions with regard to, for example, snow conditions where staff are frequently unavailable for clearing work at a majority of locations. There are, however, continual appraisals as to how best this situation can be resolved. Examples are the wider installation of point heating equipment, increased use of road/rail vehicles, redeployment of itinerant station 'heavy cleaning squads', and the involvement of private contractors equipped to carry out snow clearance from platforms.

## WEATHER WARNINGS

For weather advice purposes ScotRail is divided into seven geographical regions (Fig. 7.1). The Weather Centres in Glasgow and Aberdeen telex British Rail in London via the British Telecom link. British Rail then transfer the received message into their own communications network and it is retransmitted back to ScotRail Operations Room (Fig. 7.2). The Operations Control Room then create a file within a British Rail database and the message is subsequently transmitted to designated area operations and technical functions. This message is required to be acknowledged on a positive basis thereby completing the communications loop.

TABLE 7.1. *Principal dimensions and resources of ScotRail*

|  |  | Percentage to the rest of British Rail |
|---|---|---|
| Passenger journeys (million) | 46 | 7% |
| Passenger route miles | 1,557 | 21% |
| Freight route miles | 143 | 10% |
| Number of passenger stations | 298 | 13% |
| Number of Parcel Points | 90 | 16% |
| Number of Employees | 13,898 | 10% |

Traction and Rolling Stock

|  |  | Fleet Allocation |
|---|---|---|
| Diesel main line locomotive | — | 208 |
| Diesel shunting locomotives | — | 44 |
| Diesel multiple units (3 Car) | — | 67 |
| Diesel multiple units (2 Car) | — | 6 |
| Electric Multiple Units (3 Car) | — | 117 |
| Loco-hauled coaches (Scottish allocation) | — | 416 |

Passenger Train services

|  | No. of passenger trains run each weekday |
|---|---|
| Local services | 1460 |
| Long Distance Expresses | 239 |

Each function designated to receive such messages is responsible for initiating appropriate action. This system was introduced in January 1987 and has been a vast improvement on the previous arrangements which were verbal and referred to traditional railway administrative divisions. These boundaries have been revised to the more relevant seven ScotRail weather areas.

Similar to the rest of British Rail, the ScotRail system has many inbuilt safety features. It is fundamental that the equipment is designed to fail safe. Activation of these devices requires alternative methods of operation which, in a tightly coordinated system, invariably causes delays and some degree of disruption to train services. The following examples illustrate how adverse weather can be detected, either because of the specific nature of the indication received or from subsequent reports from the location concerned.

On all passenger routes facing points are interlocked with signalling equipment. Before signals can show a proceed aspect, agreement must be obtained with point detection equipment. The tolerance is not large and the operation of points during snow or freezing conditions can cause a loss of detection because of compacted snow or ice. The track circuit is a low voltage current which is passed between the pair of rails on a track when the passage of a train occurs. This gives a positive location of the train and operates various signalling and safety devices. When track circuits

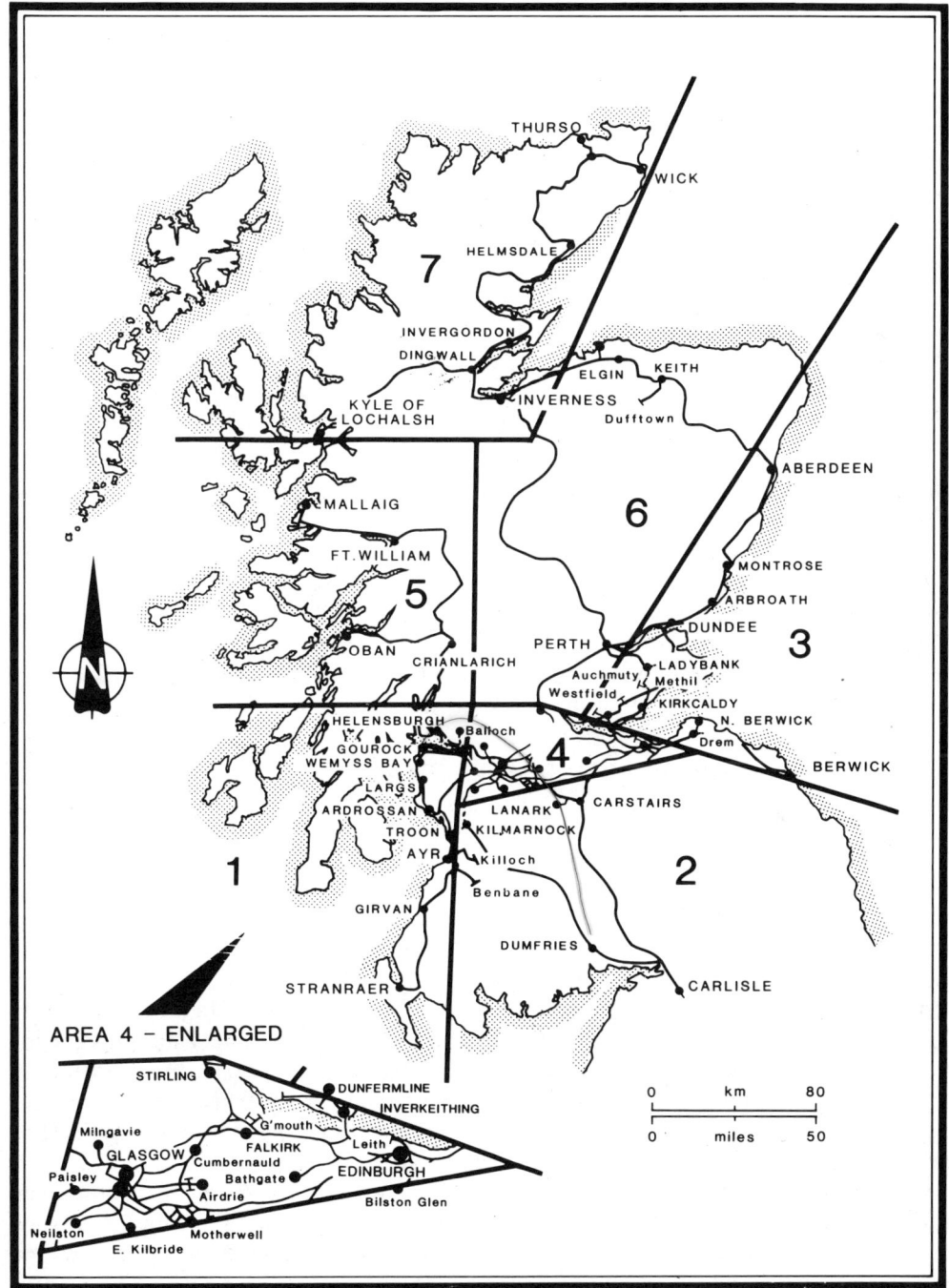

FIG. 7.1. Division of Scotland into forecasting areas for ScotRail.

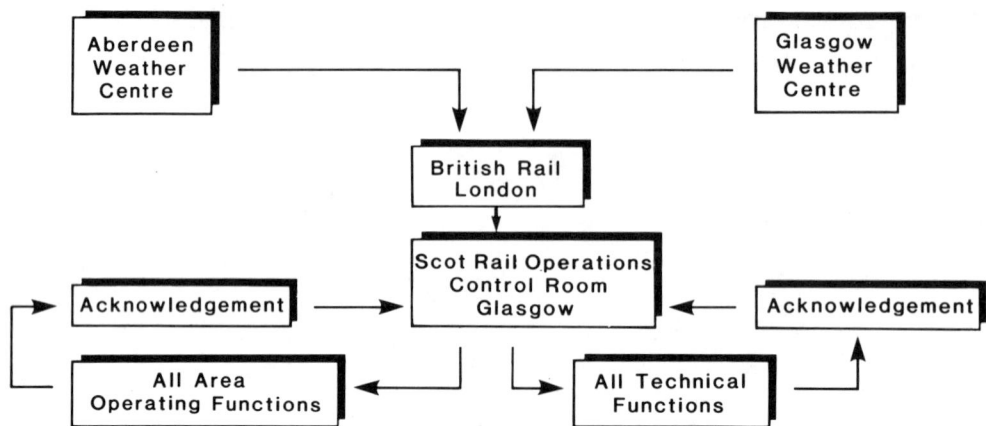

Fig. 7.2. Dissemination of weather forecasts between the Weather Centres and ScotRail.

activate without the passage of a train this may sometimes be due to flooding. Flood detection equipment is installed in selected locations such as the low level lines at Glasgow Central and the twin bore tunnels at Dalmuir.

An important feature is the quality of reports from staff on the ground. A recent development is the introduction of radios on all locomotives working the West Highland and Wick/Thurso routes in addition to the Glasgow suburban routes. There is a rolling programme to equip most of British Rail's mainline locomotives with radios in the drivers' cabs. These radios are primarily for emergency use.

## TYPES OF SEVERE WEATHER AFFECTING SCOTRAIL OPERATIONS

Flooding can weaken and possibly wash away the track bed. In extreme cases bridges and viaducts have been swept away. Normal movement of all rail traction should cease when the water level reaches a point 0·05 m below the top of the running rail. However, locomotives may run on flooded sections at walking pace, provided the level of water is no more than 0·1 m above rail level. Hot weather can cause continuous long welded rail to expand and points can go out of correspondence. Fog can retard movements, particularly in passenger and freight terminals. In the very few remaining semaphore signalling areas, additional safety margins for fog or falling snow conditions are introduced, which may have a direct effect on punctuality. In multiple-aspect colour light signalling areas there are no running restrictions as the signal aspects, as well as being highly visible, are supplemented by an audible and visual repeater system in the drivers's cab.

When sub-zero temperatures are forecast specific action is taken to protect coal traffic which is conveyed in hopper wagons, as coal will not flow freely under such conditions. Outwith very cold periods, coal trains are often loaded one day and moved the next depending upon acceptance arrangements at the discharge installations. However, when −3·0°C is expected this arrangement is suspended and, when a temperature of −5·0°C is forecast, loading is not permitted before 08·30 hours. This restriction requires a great deal of short-term planning to best meet the needs of both British Coal and the South of Scotland Electricity Board.

The formation of icicles in tunnels and underbridges can short out the 25 kv supply to overhead line equipment. Such problems are more prevalent in tunnels in rock and older bridges which have a poor degree of sealing from water penetration. When low temperatures are forecast ice patrols monitor known trouble spots and, if necessary, break off icicles with specially designed fibreglass poles.

Rolling stock design is an important factor in weather susceptibility. During extremely low temperatures toilet storage tanks are drained on older passenger coaches which are then provided with emergency water containers. Modern vehicles are more resilient due to better design. Locomotives not supplied with antifreeze are drained of water when not required or are out of traffic for repairs. The passage of trains at speed further lowers temperatures by heat loss. Modern designs of rolling stock attempt to avoid draughts where possible and the use of methylated spirit in air lines and valves reduces the risk of moisture freezing.

High winds can affect services when foreign objects are blown on to the track or overhead equipment. The latter may foul the pantographs on passing trains. Wind alarms at Carstairs, Elvanfoot and Cranberry trigger off a number of reactions in control offices. In the case of the West Coast Main Line, electrically hauled trains are reduced to 80 mph when only one pantograph is in operation, and 60 mph when two are in operation. When high winds are forecast the freight yards at Mossend and Carlisle are informed and a special check is made of wagon sheet security as these may be whipped into overhead wires. High winds may also blow heavy sea spray onto coastal routes. When this occurs, for example in the Cove Bay area, Inter-city 125 trains which, for any reason, have only one of the two power cars functioning are planned to be locomotive assisted between Aberdeen and Stonehaven. On the Glasgow–Largs line in the Saltcoats area the sea has, on exceptional occasions, affected the overhead line equipment. At this location 50 kv insulators are installed instead of the standard 25 kv specification. In order to cope with the accumulation of salt, these heavier insulators are coated with grease which facilitates cleaning. The wiring runs have been redesigned to reduce the extent of exposure to arcing under such conditions.

Snow conditions are generally considered to be the most difficult of the weather-related problems in Scotland. No other element demands the level of resources which necessarily require to be made available in order to keep routes open to traffic. Unprotected traction units cease normal movements when the depth of snow reaches 0·15 m above the top of rail level. Another major problem can be the ingress of snow to electrical traction motors which, when subject to prolonged exposure, eventually short-circuit and fail.

Approximately ten years ago it was recognised that aerial telegraph wires were inadequate because, in the more open areas, line breaks were inevitable due to the heavy weight of snow on the wires or poles which could be toppled by high winds. A programme was embarked upon to lay signalling cable underground on certain parts of the West Highland line and between Georgemas Junction and Forsinard in the far North. At the same time emergency telephones for train crews were provided at regular intervals along the route. This development satisfied the needs of the time but has itself been superseded by the introduction of Radio Electronic Token Block equipment, which was introduced on the Wick/Thurso lines during 1985. The signalman at Dingwall is in continuous radio contact with all trains under his control.

This RETB system has now been extended to the West Highland line the transmitter for which stands on Mealla Buiridgh at 1080 m above sea-level. The system has proved to be very efficient and the constant communication is valuable during bad weather which can quickly be reported by the train crew from any part of the route.

## COLD WEATHER OPERATIONS

Locomotives fitted with miniature snow ploughs work trains on vulnerable routes and are also utilised to patrol lines where snow depth is known to be 0·45 m or less. The independent snow ploughs are the traditional heavy duty stalwarts of ScotRail's battle against snow conditions. They are wheel mounted and their operational formation is normally that of one plough at each end of two coupled mainline locomotives. There are two based at Fort William, four at Inverness, two at Perth, and two at Eastfield. Snow ploughing places a heavy burden on locomotive provision, each pair of ploughs requiring two, both of which need to be replaced after 8 hours of hard ploughing. Buffer snow ploughs are still regarded as experimental and are intended only for patrol work and for ploughing drifts up to 2 m high.

The Beilhack patrol plough, imported from Germany, is wheel mounted and is also used for patrol work and 2 m drifts. They may be attached to either diesel or electric locomotives. The normal position of the blade is in a side position so that the snow is pushed or rolled onto the left hand side of the track in the direction of travel. This does cause some problems on platforms. The Beihack self-propelled rotary snow blower can run to any site at up to 40 mph when hauling its own support vehicle. It clears a 2·7 m path and can blow snow 30 m clear of the line. In ideal conditions 5000 tonnes of snow can be blown each hour. Progress depends on the depth of snow which at 2 m will allow a 3 mph movement and at 3·5 to 4·3 m, 0·5 mph. The machine was first used in earnest during the early part of 1984 on the Inverness to Wick line. As a result, the line was reopened in three days. This compares with similar conditions during 1978 when it took two weeks to reopen.

ScotRail's Civil Engineer has fitted some track maintenance machines with snow ploughs. It is intended that these modified self-propelled machines will be used for patrol work and assisting with yard clearance operations. In addition, portable lances have been made to operate from the air-supply available on mainline and shunting locomotives. These are intended for clearance of snow and ice from points and crossings located on running lines and within yard installations.

On the running lines and loops in the ScotRail system there are approximately

2800 point ends. Of these, approximately 2500 are heated, normally by electricity but occasionally by propane gas. Electric heaters of the 'Pad' type have now been superseded by the more efficient 'Strip' heaters each metre of which requires 200 W of power. Thus a typical point-end, requiring four strip heaters each six metres long, consumes 4800 W of electricity. This equipment is activated by 'Findlay-Irvine Icelert' sensors, which are set to operate when 3°C is reached during snow conditions and 0°C outwith snow conditons.

## TRAIN SERVICE CONTINGENCY PLANNING

The ScotRail Regional Operations Control Room is located in Glasgow and is the focal point of all current railway operations within Scotland. The primary role of Operations Control is to maintain the train plan and, in the event of a disruption, intervene and direct contingency arrangements. The Traction and Resources sub-section is responsible for meeting actual current movement commitments in terms of provision and manipulation of locomotives for all trains, and of train crews primarily for freight services. The Train Running sub-section comprises several staff each responsible for a geographical area of route. The size of each section is determined by the level of activity within it. Staff within the sub-section are particularly involved with the real-time emergencies which arise. Rapid preventative decisions are required to avoid any situation worsening more than necessary.

After routes have been protected and defended as far as practicable, there may be occasions when it is simply not possible to continue to do so. In these circumstances, rather than spread route clearance resources too thinly, efforts are concentrated on keeping selected routes and sidings open, as previously determined within ScotRail in the light of commercial priority. Movement of fuel for industry and domestic use would, for example, be essential in a major crisis. Area Managers have examined track layouts and have established which specific tracks can be used to best advantage using as few switching movements as possible.

The ScotRail timetable contains numerous through trains which start from Inverness or Aberdeen and form part of the core service from Edinburgh to the south. Inter-city 125 train sets are intensively diagrammed and, to allow a maximum utilisation, the turn-round time at terminals may be as little as 45 minutes after a 550 mile journey. During adverse weather, which may only be at the opposite ends of a lengthy route, such tight turn-round times are quickly eroded and, to avoid perpetual late running and cancellations, the Operations Control Office manipulate the service for best overall effect. A careful watch has to be kept on available locomotives and rolling stock which must still receive proper maintenance and fuelling. An ever-growing computer system is indispensable to efficient operation.

In conclusion it must be said that most of the inherent difficulties which exist within a variety of fields have been faced and either overcome, or found to be too expensive or insoluble. The degree of weather protection is determined by money, and by judgement as to how much investment can be afforded for normal circumstances of bad weather compared with relatively infrequent exceptional circumstances.

# SECTION THREE

# AGRICULTURE, WATER AND WIND RESOURCES

# Services to agriculture: past, present and potential

B. A. CALLANDER

*Meteorological Office, Bracknell*

## INTRODUCTION

THE Isle of Tiree, deprived by the wind of any trees, is known as the 'land of the low-lying barley'. In spite of this, Tiree was traditionally the bread-basket of the Hebrides. Such is an example of the accommodation that farming can engineer with the local climate. Scotland, as a whole subject to a more testing climate than the rest of Britain, has developed its agriculture to cope with the limitations, and sometimes to exploit the advantages, of its situation. The danger of farming within such an environment is that difficult weather is accepted as an unchangeable variable, precluding consideration of the possibility that the impact of the weather may be further reduced through the judicious use of better weather and climatological information.

## THE CLIMATIC CONTEXT OF SCOTTISH AGRICULTURE

A review of the Scottish climate was submitted to a Select Committee on Scottish Affairs looking at land resource use (Meteorological Office, 1971). More recently, Francis (1981) described the climate of the agricultural areas of Scotland, subdivided into 15 climatological areas. It is appropriate to summarise here some of the main features.

Substantial areas in Scotland, including the western and northern Isles, a wide coastal belt in SW Scotland and extensive areas on the eastern side of the country, receive solar radiation totals similar to those recorded northwards of Hull and Blackpool. However, monthly totals in East Anglia and much of the South of England are uniformly higher than in any part of Scotland. Perhaps more surprisingly, for a given latitude, annual radiant income in Scotland is generally less than in southern Scandinavia. Only during the sunniest period, April to July, is radiant income similar to most other areas of England and Wales.

Annual average rainfall varies from around 1000 mm in the wetter agricultural areas of the west to around 700 mm for much of the east coast from Eyemouth up to Montrose and on the coast around the Moray Firth. Inter-annual variability can be large. For Scotland as a whole the yearly mean is about 1400 mm, but can vary between 1000 and 1600 mm. On at least one occasion both extremes have been attained in adjacent years: 1954—1626 mm and 1955—1041 mm. Spring and early summer contain the driest months; July then sets a pattern of higher rainfall which continues to the following January.

With such copious rainfall, irrigation is not normally required for agricultural production. Harper (1974) examined the long-term irrigation need for Scotland and concluded that "only a very limited area of Scotland, entirely on the East Coast and mainly on the southern coastal strips of the Moray Firth and the Firth of Forth, has a total irrigation need in 20 years in excess of 2500 mm, and a need in the driest years in the range 250 to 330 mm". In comparison, large agricultural areas of England have this requirement. Only for intensively grazed pasture or high value horticultural crops in areas with an average annual rainfall around 750 mm is irrigation generally a serious consideration.

Temperatures in agricultural areas are generally moderated by the surrounding sea and, apart from a few inland eastern areas, severe frosts represent one of the lesser hazards to Scottish agriculture. Accumulated temperature in the important January to June period averages from 1380 degree days in Ayrshire to only 990 degree days in inland parts of the north east. In comparison, January to June accumulated temperatures for East Anglia and Lincolnshire generally exceed 1400 degree days and, in coastal areas, often exceed 1500 degree days.

The traditional starting period for the growing season is when average daily air temperature reaches and remains above 5·6°C. The milder west coast reaches that point in late February or early March, similar to much of central and southern England, but in eastern coasts of Scotland dates are more typically mid-to late-March. The reduction in length of growing season with height is particularly severe in Scotland: as a general rule six days are lost for every additional 100 m of altitude, but in Scotland the figure is 14 days per 100 m, which is "more rapid than in any other temperate land for which we have data" (Manley, 1970). As is the case for rainfall, growing season length can exhibit considerable interannual variability: for Eskdalemuir in the west Gloyne (1968) found a mean GSL of 205 days and standard deviation of 20 days. The maximum was 266 days and the minimum 174 days. In the agricultural areas of the west, the growing season averages 250 days but, in the east, the average length is almost a month shorter at about 220 days (Francis, 1981).

World maps of windiness show that northern Britain experiences mean annual windspeeds at low levels equalled by few populated land areas (Gloyne, 1971). In terms of horticulture, it has been shown that shelter is necessary if production of soft fruit is to be maximised (Waister, 1971). The effect of horizontal habit on the yield of Tiree's barley has not been documented.

## THE PAST

Scotland has tended to lag behind England in the provision of meteorological services for the benefit of agriculture. In 1951 the Department of Agriculture for Scotland (DAS, now DAFS) approached the Air Ministry with a request to set up a specialised Agricultural Meteorology Unit (AMU) in Edinburgh, similar to those already operating since the 1940s in the regions of the Agricultural Development and Advisory Service (ADAS) in England. Such a unit was established in 1952. In 1971 a Select Committee on Scottish Affairs (Meteorological Office, 1971) commented: "Meteorological information is important for many planning decisions connected with physical development, agriculture, forestry and water resources. It has been evident for some time that information on meteorological characteristics in Scotland is lacking and compares unfavourably with that obtained in England. Consideration is at present being given to the extension of weather station networks in Scotland."

Due to financial pressures the Agricultural Meteorology Unit in Edinburgh closed on 31 December 1981: Responsibility for the provision of distinctly agricultural climatic enquiries officially devolved to Bracknell headquarters, though enquiries are usually first directed to Edinburgh Met Office. Without day-to-day involvement of the Met Office in Scottish agriculture, the number of enquiries quickly dropped. This does not necessarily indicate a lack of need, since there are many examples from both within and outside agriculture where potentially valuable advice is not sought simply because the means of obtaining it are not convenient.

## THE PRESENT

Edinburgh Met Office handles the purely climatological enquiries (see Tabony and Brown, this volume) but, for more specialised advice, reference is made to the Agriculture Section at Bracknell. Many years of collaboration with ADAS have produced a wide range of computer programs that use the extensive climatological data base at Bracknell to produce statistics of direct relevance to agriculture. Examples are spray-day analyses, evaporation climatology and glasshouse heating requirements.

A number of organisations in Scotland, including the Scottish Agricultural Colleges (SACs), receive data direct from Bracknell. In practice SACs are the major players in the application of meteorology to agriculture. Through the East of Scotland College (ESCA) the three colleges receive daily and weekly bulletins of meteorological information and agromet products which include weather-related plant disease indices. The data are bought strictly for internal use by the colleges, in the sense that commercial services can be based on the data, but the data themselves cannot be sold or passed to a third party. However permission is usually given for a third party to access these data if it is for the purposes of research only.

The source of these data is the network of synoptic and daily reporting climate stations in Scotland. Fig. 8.1 shows the location of these stations. They represent the maximum current number of stations from which real-time data are available. These stations are by no means the only locations where meteorological observations are carried out, and Fig. 8.2. shows the denser network of climatological, agrometeorological and sunshine stations. Rainfall is measured on an even denser network (see Singleton this volume). Observations from these stations are reported to Bracknell each month, so are only available for use some weeks after the data have been collected. Many of the observations are made at sites run by the SACs and the Scottish Agricultural Research Institutes (SARIs) and a system of data exchange occurs naturally between these bodies if the need for real-time data arises.

FIG. 8.1. Climatological data bank stations providing hourly synoptic observations and daily climate information in Scotland.

Services to agriculture: past, present and potential

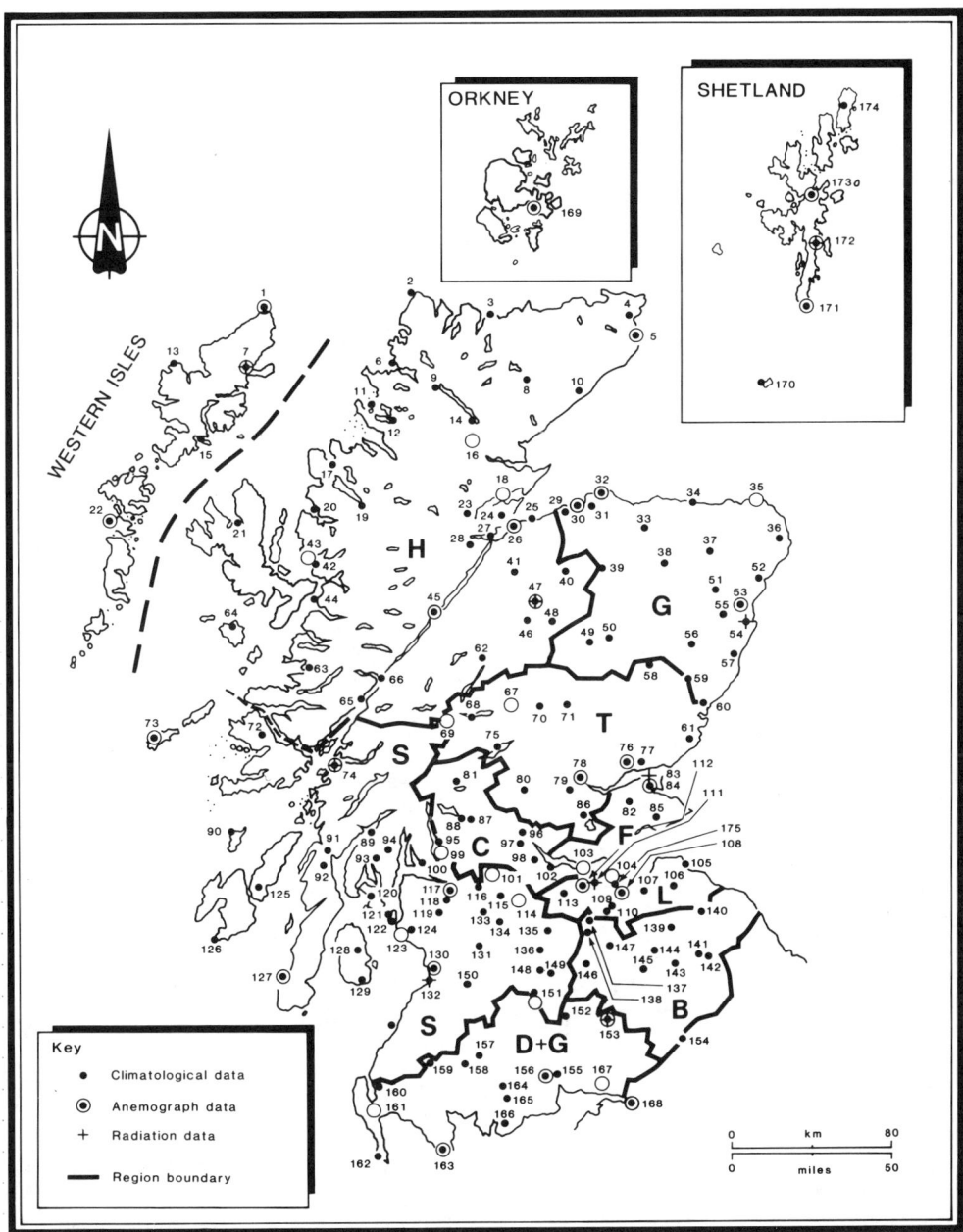

FIG. 8.2. Meteorological stations for which data are published in the *Monthly Weather Report*.

*Key to Stations in Fig. 8.2.*

1. Butt of Lewis
2. Cape Wrath
3. Torrisdale
4. Keiss
5. Wick
6. Duartmore Bridge
7. Stornoway
8. Kinbrace
9. Cassley
10. Berriedale
11. Inverpolly
12. Knockanrock
13. Knoc
14. Lairg
15. Scalpay
16. Shin
17. Poolewe
18. Invergordon Harbour
19. Kinlochewe
20. Diabaig
21. Prabost
22. Benbecula
23. Dingwall
24. Fortrose
25. Nair
26. Dalcross
27. Inverness
28. Lentran
29. Forres
30. Knoloss
31. Aldroughty
32. Lossiemouth
33. Keith
34. Banff
36. Forehill
37. Fyvie Castle
38. Gartley
39. Glenlivet
40. Grantown-on-Spey
41. Tomatin
42. Plockton
43. Duirnish
44. Eilanreach
45. Fort Augustus
46. Lagganlia
47. Aviemore
48. Cairngorm
49. Braemar
50. Balmoral
51. Inverurie
52. Culterty
53. Dyce
54. Aberdeen
55. Craibstone
56. Finzean
57. Stonehaven
58. Whitehillocks
59. Fettercairn
60. Montrose
61. Arbroath
62. Dalwhinnie
63. Inverailort
64. Isle of Rhum
65. Onich
66. Fort William
67. Tummel Bridge
68. Dall
69. Rannoch
70. Faskally
71. Kindrogan
72. Aros
73. Tiree
74. Dunstaffnage
75. Ardtalnaig
76. Mylnefield
77. Dundee
78. Perth
79. Strathallan
80. Drummond Castle
81. Balquidder
82. Cupar
83. Shanwell
84. Leuchars
85. Belliston
86. Kinross
87. Loch Venachar
88. Aberfoyle
89. Glenbranter
90. Colonsay
91. Lochgilphead
92. Stronachluin
93. Benmore
94. Ardentinny
95. Arrochymore
96. Parkhead
97. Stirling
98. Falkirk
99. Blairlinnans
100. Helensburgh
101. Inchterf
102. Grangemouth
103. Forth Road Bridge
104. Leith Harbour
105. Dunbar
106. Nunraw Abbey
107. Pathhead
108. Royal Observatory
109. Bush house
110. Penicuik
111. East Craigs
112. Turnhouse Airport
113. Livingston
114. Salsburgh
115. Coatbridge
116. Glasgow
117. Abbotsinch
118. Paisley
119. Uplawmoor
120. Rothesay
121. Millport
122. Hunterston
123. Ardrossan
124. Dalry
125. Craighouse
126. Upper Killeyan
127. Machrihanish
128. Brodick Castle
129. Kildonan
130. Prestwick
131. Drumclog
132. Auchencruive
133. East Kilbridge
134. Strathclyde Park
135. Carnwath
136. Lanark
137. West Linton
138. Blyth Bridge
139. Blythe
140. Whitchester
141. Floors Castle
142. Kelso
143. Greycrook
144. Galashiels
145. Bowhill
146. Stanhope
147. Glentress
148. Abington
149. Camps Reservoir
150. Cumnock
151. Leadhills
152. Auchen
153. Eskdalemuir
154. Gelder Castle
155. Dumfries
156. Drungans
157. Glenlee
158. Clatteringshaws
159. Bargrennan
160. Penwhirn
161. West Freuch
162. Mull of Galloway
163. Whithorn
164. Glenlochar
165. Threave
166. Girdstingwood
167. Chapelcross
168. Carlisle
169. Kirkwall
170. Fair Isle
171. Sumburgh
172. Lerwick
173. Selaness
174. Baltasound
175. Royal Botanic Gardens, Edinburgh

*Services to agriculture: past, present and potential*

An added stimulus for the SACs and SARIs to exchange meteorological information is the acknowledged lack of real-time Met Office stations in the agriculturally important areas of east Scotland. There is no short-term solution to this problem, but the Met Office is considering ways in which the availability of real-time data can be improved. The Agriculture Section of the Office can urge the siting of new automatic weather stations (AWSs) in areas of agricultural interest but, under the present economic climate, the rate of appearance of new stations will be low. As an alternative, the possibility of setting up a mechanism for private AWSs to feed data into the Met Office is being studied.

Forecasts are provided through a variety of channels including radio, TV, newspapers and Videotext. The more specialised forecasts for farming appear in the Sunday lunchtime Farming TV programme and on Prestel's Farmlink service. Weathercall, operated by Telephone Information Services in conjunction with the Met Office, provides both regional 24 hr and national 5-day forecasts, though these are not aimed specifically at agriculture. Apart from open Prestel, agricultural users have access, on payment, to the Farmlink closed user group (CUG) within Prestel; Farmlink has now absorbed the other agricultural CUGs Uniform (National Farmers Union) and Agviser (ICI). In addition to the weather forecasts already mentioned, daily forecasts of accumulated temperature for 7-days ahead, aimed at guiding the timing of application of fertilizers to grassland, appear on Farmlink in the Spring. The same forecasts are published in *Farmers Weekly*. This service is sponsored by UKF Fertilisers Ltd who, through an editorial column in FW add advice, warning and interpretation of the forecasts where necessary. The Met Office also operates a nationwide rainfall and evaporation calculation system called MORECS. Using data from the network of Fig. 8.1, rainfall, evaporation, soil moisture deficit and hydrologically-effective rainfall are presented as areal averages for 40 km squares. Limited MORECS information is available nationwide on Farmlink viewdata service for the period April to October.

For even more precision and flexibility, individuals can take out a consultancy with the local Weather Centre, giving access to the duty forecaster on an ex-directory number any time of day or night. The cost of these consultancies is particularly competitive when set against the potential savings. One potato farmer in East Anglia estimated that his consultancy service saved him 5% of his spray bill, equivalent in his case to over 16 times the annual consultancy cost. However, a study in 1984/85 found that only a small number of farmers in Scotland subscribed to such a service with either Glasgow or Aberdeen Weather Centres. The vast majority apparently obtained their weather information through the media. These comprise mainly radio and TV forecasts and Weatherline (now superseded by Weathercall) which are not specifically designed for their needs and will not give precise details of expected timing or geographical variation.

## THE FUTURE

A survey of the potential benefits to agriculture of a Scottish radar network has estimated potential savings of around £650K. The breakdown of this sum is shown in

TABLE 8.1. *Estimated annual potential savings for agriculture from weather radar in Scotland*

|  | £ |
|---|---|
| Crop spraying | 400,000 |
| Hay and silage making | 137,000 |
| Irrigation | 17,000 |
| Cultivations (cereals) | 50,000 |
| Soft fruit harvesting | 45,000 |
|  | Total 649,000 |

Table 8.1. A major omission is any sum against harvesting, which was not quantified in the study. With the relative wetness of the autumn period in Scotland, the ability to time precisely the arrival of a shower or rainband must yield considerable savings of time and money. It is not yet known how soon radar will be available in Scotland, and there are other ways in which Scottish agriculture could better exploit information available from the Met Office. This is in the field of agricultural models, particularly those designed to aid operational decision-making.

Meteorological information is rarely used directly as advice. For example, in irrigation calculations rainfall is a very important variable, but the decision to irrigate depends on soil moisture deficit, not on rainfall. Agriculture is dependent on the weather, so many models employed or under development in agriculture such as irrigation scheduling, work planning, crop growth and development, market trend forecasting or energy use in crop stores or glasshouses, must inevitably use meteorological data as an input. Important models are being developed in Scotland. For example, the potato growth model developed at the Scottish Crops Research Establishment is regarded as the one of the best in the world. Unfortunately, few organisations involved in developing or exploiting these models regard the Met Office as more than a passive supplier of data. The Met Office is willing to supply historical or current data to any customer, but the commercial value of many models is greatly increased if forecast weather information is incorporated. These data are not generally available outside the Met Office.

The Met Office Agriculture Section at Brackwell is actively developing the incorporation of numerical forecast information with agrometeorological models. While it can normally identify those models or agricultural activities that could benefit most from real-time or forecast weather information, the Met Office does not have the specialised farming knowledge or the direct contacts with farmers that would allow the development of new products. As a consequence, progress in specific areas has been achieved through cooperation with agriculturally-based organisations. Examples are the UKF T-SUM forecasts; the ADAS 'Irriguide' service which is the only UK irrigation scheduling service to incorporate forecast rainfall and evaporation; and forecasts of potato blight weather, trials of which were

## Services to agriculture: past, present and potential

run in conjunction with ICI in 1987. So far, Met Office relationships with Scottish agriculture in this developing field fall in the 'passive' category.

A range of weather services are available to Scottish agriculture, but the exploitation of these services by the agricultural community remains minimal. The lesser emphasis on weather and climate information in Scotland compared to England inevitably results in less interaction between the user and the Met Office, less opportunity to develop specialised services and a weaker influence in the improvement of the observing network in the agricultural areas of Scotland. It is hoped that in future the relationship between Scottish agriculture and the Met Office might attain a higher profile than that of Tiree's low-lying barley.

# Water resource management and flooding

R. J. SARGENT

*Forth River Purification Board, Edinburgh*

## INTRODUCTION

Like many other industries, the water industry is affected by weather extremes, particularly precipitation extremes, at both ends of the scale. Too little rainfall, and water resources are depleted and polluted by rivers receiving effluents with insufficient dilution. Too much rainfall, and flooding and its attendant problems of flood prediction and prevention result.

This paper deals with the sensitivity of the Scottish water industry to both types of extreme condition. In particular, the sensitivity of water resources and river quality to drought are discussed and, at the other extreme, data on the prevalence and costs of flooding in Scotland are presented. Ways in which flooding is mitigated are also discussed.

## SENSITIVITY TO DROUGHT

The sensitivity of different water resource schemes to drought can vary greatly. This arises from a combination of the different sources which can be used and the engineering design of the scheme. Three main sources are used in the UK, direct abstraction from a river, impoundment, either of runoff from a catchment or of water pumped from another source, and groundwater. Sometimes more than one source is used conjunctively, often to ameliorate the effects of drought. The sensitivity of each of these to drought depends on the reliability of the supply and the storage available. The bulk of Scottish supplies comes from upland impoundments, with a small amount from direct river abstraction. Both of these can be sensitive to short, intense drought. Groundwater supplies are usually less affected by such occurrences, but they are under-developed in Scotland, accounting for no more than three percent of supplies (Robins, 1987).

The sensitivity of water supplies also depends upon their design. Typically, the upland impoundments in Scotland are underdesigned, that is they do not catch all the runoff of the catchment and they also spill significant quantities through the winter period. This means they do not enter the summer period with as much water as they could, and supplies can be very sensitive to spring/summer droughts. Conversely, supplies in England are often from large reservoirs or groundwater which rely on winter rain to replenish supplies. These supplies are, therefore, more sensitive to winter droughts. Thus we see the 1984 drought, in which parts of Scotland during the period April to August 1984 received less than 30% of normal rainfall (Marsh and Lees, 1985), caused great difficulties for supplies in Scotland, particularly those relying on upland reservoirs such as the Carron in Central Region and several in Strathclyde Region. Supplies in England were less seriously affected, though in areas with upland reservoirs such as the Lake District, Pennines and Wales, supplies were also badly hit.

The 1976 drought, on the other hand, was much longer and affected the winter replenishment of 1975/76. Supplies in the south of England were more affected on that occasion. It must be acknowledged that these droughts were not distributed uniformly and the 1976 drought was more severe in the south of the country, but these two examples do demonstrate the varying sensitivites of water supply schemes, and the vulnerability of Scottish supplies to drought.

A further problem associated with drought is the effect of low river flow on water quality. Effluents are normally permitted to be discharged on the basis of the river's ability to purify the organic content of the discharge by natural processes. These processes consume oxygen, which is dissolved in the water, and if overloaded the water can become deoxygenated and thus unable to support aquatic life. Consent to discharge organic wastes to watercourses is granted by River Purification Boards with this limit in mind, and usually a design low flow is used to match the quantity and quality of effluent with the river's ability to purify it. During a drought, the river discharge can fall below this design flow and thus water quality becomes threatened. This problem is exacerbated if the drought occurs during summer months when the biological purifying processes operate faster and the water naturally holds less dissolved oxygen at its saturated level.

In some rivers the flow during a drought can be substantially made up from water discharged by water supply reservoirs via a "compensation flow" agreement. If effluents are discharged to these watercourses they are particularly vulnerable since drought can mean a reduction in their flow, not only from natural means but also by a reduction in the compensation flow. This can only be done via an Emergency Water Order laid before the Secretary of State, but, if granted, it can have a significant effect on the receiving water.

As an example, the Forth at Stirling is particularly vulnerable to compensation flow reduction. During drought conditions its flow is dominated by discharges from the Loch Venechar compensation reservoir in the Trossachs, supplying a compensation flow for the Loch Katrine supply to Glasgow. In the upper Forth estuary the river receives organic wastes from a number of large sources, including Stirling sewage treatment works, and yeast factory waste at Alloa. The consents held by these dischargers are based on the freshwater flow to be expected in the river, including the compensation flow. In the 1984 drought, this compensation flow formed 80% of the freshwater flow in the river at Stirling, where the dissolved oxygen fell to 5% of saturation (Forth RPB, 1984). The drought severely depleted water resources in the Katrine reservoir and Strathclyde Regional Council were forced to apply for an Emergency Water Order towards the end of the drought. This would have drastically reduced the freshwater flow into the upper estuary and seriously damaged water quality. Fortunately, the drought broke before the reduction could be brought into effect but the vulnerability of the estuary was amply demonstrated and steps have since been taken to lessen the impact of a repetition.

R. J. Sargent

# SENSITIVITY TO FLOOD DAMAGE IN SCOTLAND

Very little work has been done on the extent of flood damage in Scotland, and the subject is much neglected compared to that in England and Wales. There is a general belief that the *per capita* cost is much less, generally because flood plain development is less intense as a result of smaller, more defined flood plains and less pressure on land for development. Nevertheless, there are cases of considerable flood damage occurring in Scotland. Table 9.1 shows a selection of these, as reported by River Purification Boards (private communications). These costs are estimates of insurance claims, and do not include the costs of most emergency operations undertaken during the flood emergencies and the general disruption to every-day life.

These flood events were all localised, affecting at most 50 houses or so. There is also evidence of more widespread flooding, though occurring much less frequently. One of the largest of these floods was the event covering south-east Scotland on 12th and 13th August 1948. This flood, caused by up to 150 mm of rain falling on the 12th alone, affected many rivers in the Lothians and Borders. Dozens of bridges were destroyed, farmland was buried under deposits of boulders and houses plastered with mud. The total cost of such damage is difficult to estimate. An attempt at the time put the cost at "a million pounds at least in Berwickshire alone" (Learmonth, 1950), which is £12·4 m in present-day terms. A total of £25 million for the affected area would probably be an underestimate, and flooding today would cost more, if only because houses and shops contain more valuable goods.

In this paper, an attempt has been made to quantify some of the annual flood losses in Scotland. This uses information obtained by the working party which was established to investigate the benefits of extending the weather radar network to Scotland. An earlier working party, reporting on the situation in England and Wales, had concluded that savings in flood damage produced by more efficient warnings represented one of the most significant benefits of weather radar in those countries. The Scottish working party had therefore to investigate the potential savings in Scotland.

Since flood warning is a function of River Purification Boards in Scotland, those

TABLE 9.1. *Selected instances of high flood damage in Scotland*

| River | Flooded area | Year | Return period (yr) | Damage (£) |
|---|---|---|---|---|
| Nith | Whitesands, Dumfries | 1977 | 10 | 282,000 |
| Nith | Whitesands, Dumfries | 1982 | 12 | 250,000 |
| Braid Burn | Cameron Toll | 1983 | 30 | 750,000 |
| Water of Leith | Edinburgh | 1984 | 50 | 160,000 |
| White Cart | Langside, Glasgow | 1984 | 20 | 900,000 |
| Conon | Conon Valley | 1984 | 25 | 500,000 |

bodies were contacted via the River Purification Boards Association for information on flood prone property. Such information was scarce, and in general covered only areas where a flood warning scheme was operated or proposed. Whilst these areas generally represent those most seriously affected by flooding, isolated areas are usually not covered by such schemes and could, in aggregate, outnumber them. The proportion of total flood damage represented by such isolated property is not known. The questionnaires sent to the River Purification Boards asked for information on the number and type of properties flooded and the return period at which flooding occurs. Standard costs of flood damage (Penning-Rowsell and Chatterton, 1977) were used to estimate the total cost of flooding in each of the areas, and the annual cost was calculated using the given return period. Some assumptions were necessary to obtain the figures, though in some areas detailed costs for exact flood return periods were supplied. Table 9.2 gives an analysis of these returns.

The annual cost of flooding of property given by this survey amounts to £54,000 (1983 costs, roughly £67,500 in 1988 money). This must be a considerable under-estimate, largely because only property in areas considered for flood warning

TABLE 9.2. *Analysis of flood damage reported by River Purification Boards (1983 data)*

| Board | Flood prone property | Return period (yr) | Annual cost (£) |
|---|---|---|---|
| Tweed | 24 houses in 6 towns | 50 | |
| | 12 houses in 4 villages | 50 | 504 |
| Clyde | 200 houses @ Langside | 20 | 7,000 |
| North-east | 50 houses and 50 shops @ Elgin | 20 | 5,912 |
| | 50 houses and 50 caravans, @ Badenoch and Aviemore | 20 | 2,375 |
| | 50 houses @ Garmouth & Kingston | 20 | 1,750 |
| Solway | Whitesands, Dumfries | 10 | 25,000 |
| | 12 houses @ Port Patrick | 25 | 336 |
| Highland | 10 houses @ Findhorn | 25 | 280 |
| | 10 houses @ Dingwall | 5 | 1,400 |
| | 5 houses @ Strathoykel | 5 | 700 |
| | 35 houses & 1 shop in Conon valley | 25 | 1,047 |
| | 12 houses @ Gairlochyt | 25 | 56 |
| Tay | 50 houses @ Perth | 50 | 700 |
| Forth | 6 yachts @ Cramond | 10 | 150 |
| | 50 houses & 6 shops @ Edinburgh | 50 | 900 |
| | 5 houses @ Bridge of Allan | 25 | 140 |
| | shops & cars @ Cameron Toll | 50 | 5,000 |
| | 51 houses, 12 shops, 3 works, 1 school & 1 hotel @ Haddington | 50 | 900 |
| | | Total annual cost | £54,150 |

was covered. Even if the total is three times this much, the true cost is very much less than in England and Wales, as originally suspected. The evidence suggests, therefore, that flooding is not a serious national problem, but there are not infrequent cases of considerable damage being caused by flooding in localised areas, such as those listed in Table 9.1.

There are two possible responses to such flooding. Either flood prevention work is undertaken, to prevent flooding of valuable property, or flood warning schemes are installed to allow suitable precautions to be taken and thereby minimise flood damage. Flood prevention schemes usually involve civil engineering projects such as channel diversion, flood bank construction and so on. These are expensive to install and often require regular maintenance. The localised nature of flooding in Scotland can make flood prevention schemes cost-effective in certain circumstances but, in general, they are too expensive to be justified by the relatively small savings to be obtained in flood damage reduction. Flood warning on the other hand can be quite cheap, and is often more suited to the small savings to be obtained over a number of restricted flood plains in Scotland. There are a number of schemes operated in Scotland, and most rely on an input of accurate meteorological data for their effectiveness.

Meteorological data are especially important for flood warning schemes in Scotland where many catchments are small and steep and the time between rain falling and flooding occurring is quite short. In these conditions a quantified warning of heavy rain to come is required; waiting until the rain has fallen often reduces the time available below the minimum required to mobilise emergency services. It is also important to realise that the sensitivity of flood prone areas to extreme weather conditions varies. Some will flood following a few hours of heavy rain, others require days of moderate rain, and still others are sensitive to very specific synoptic conditions. The antecedent conditions in the catchment are also important, and the rate of rainfall is only one factor governing the likelihood of flooding.

The meteorological data input must, therefore, be tailored to the needs of the scheme. A specific example of this is the flood warning scheme covering Haddington in East Lothian, one of the areas most affected by the floods of 1948, but also subject to more regular if less spectacular inundation. Analysis of the previous floods at Haddington revealed that the most serious were caused by very specific meteorological conditions. In each case a slow moving depression had been centred over the North Sea, with a near stationary occluded front present over East Lothian. Under these conditions a north easterly airflow causes an orographic enhancement to rainfall over the Lammermuir Hills, which persists for several hours. The flood warning scheme is usually triggered by the Glasgow Weather Centre who look for the possibility of these, or similar conditions, occurring over the area. A secondary system is also installed (Sargent, 1985), which automatically monitors rainfall and makes estimates of runoff using a mathematical model of the catchment. Though more precise in its forecasts of runoff, this system provides less warning time (about 4 hours) and so forms a useful backup to the primary alert system.

Forecasts of both rainfall and runoff could be much improved if weather radar

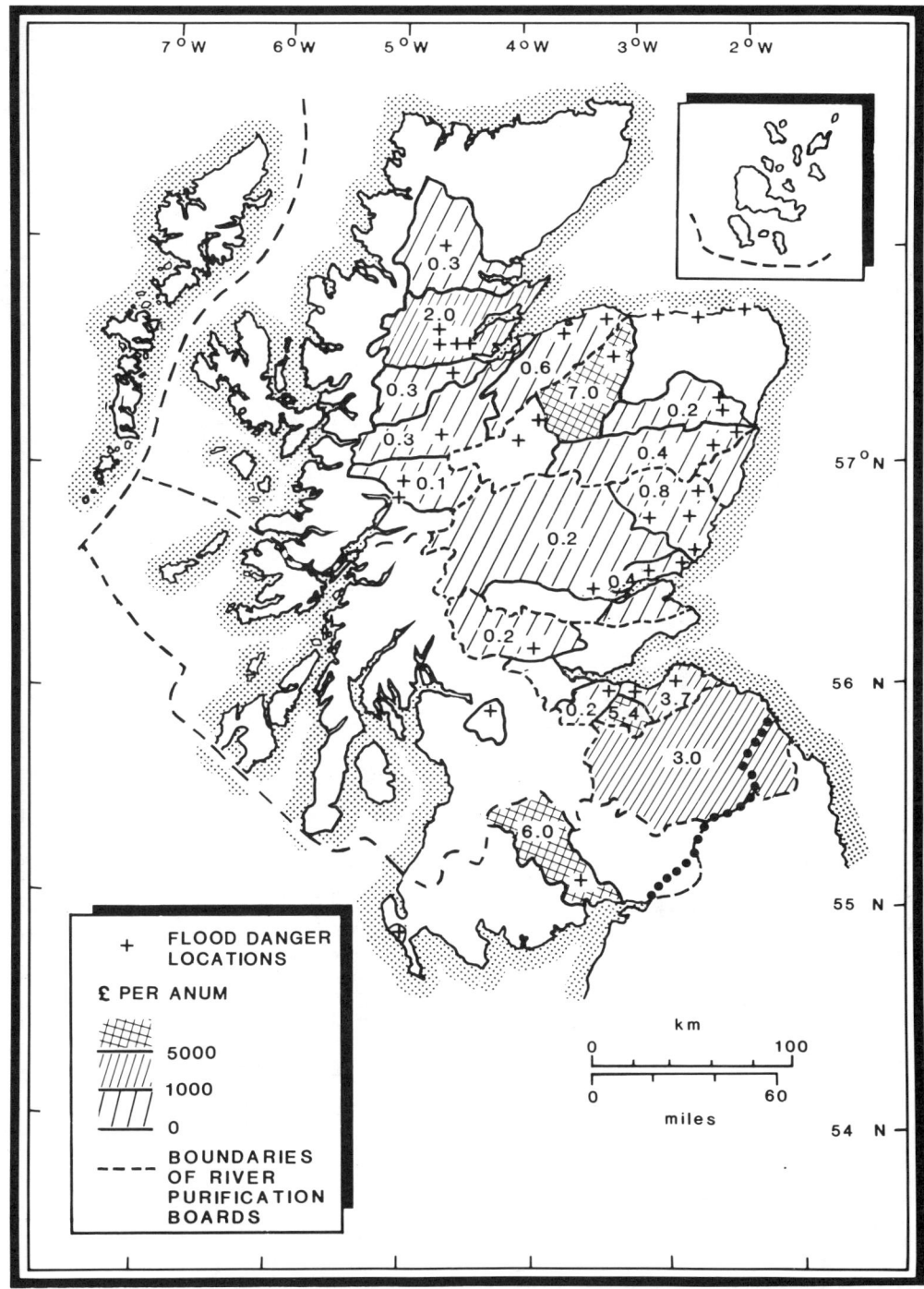

Fig. 9.1. Estimated monetary benefits of weather radar to flood warning in Scotland: precipitation measurement only.

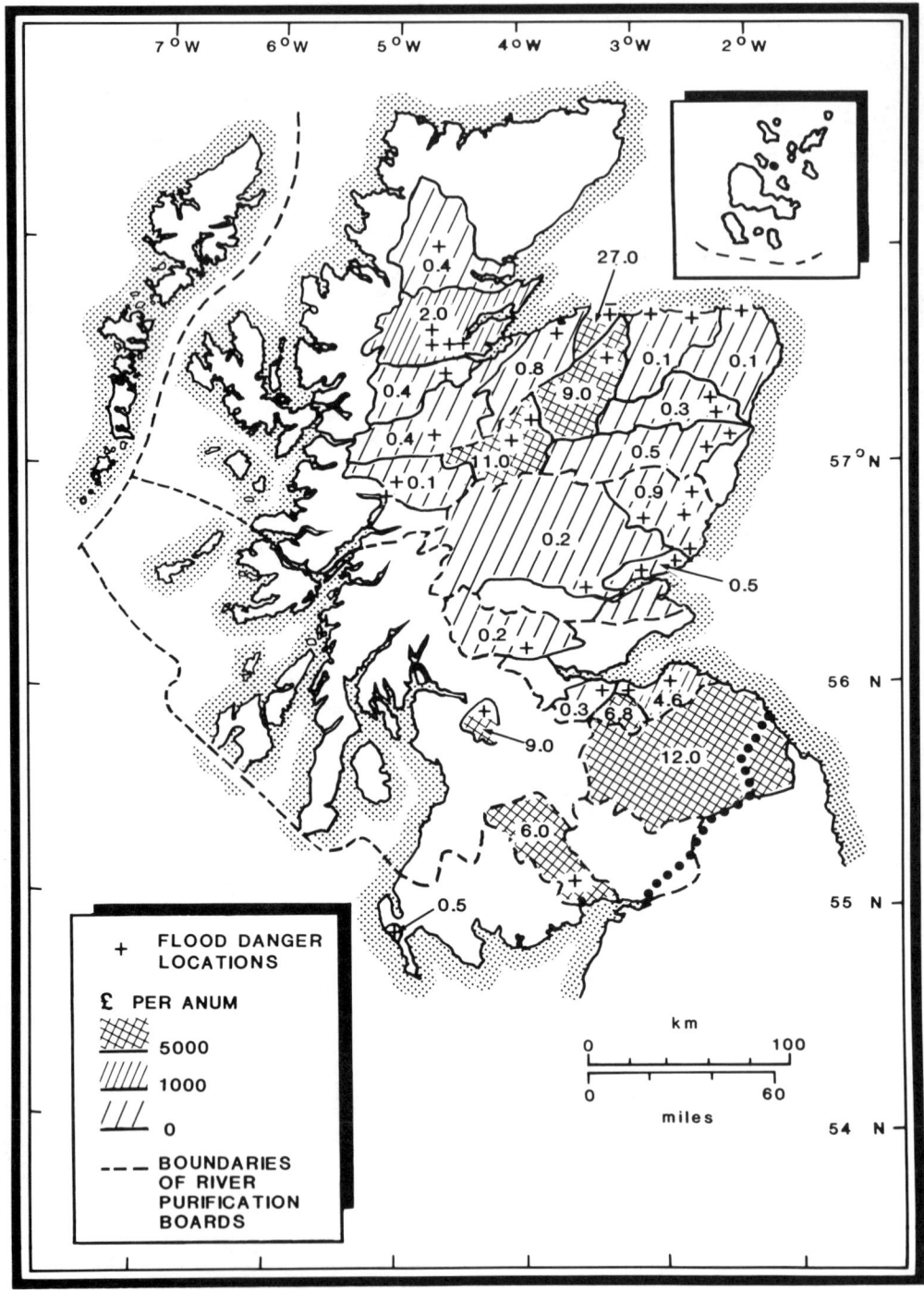

FIG. 9.2. Estimated monetary benefits of weather radar to flood warning in Scotland: 0–4 hour precipitation forecasts.

information were available. Weather radar networks, such as those established in England and other countries, enable the rate of rainfall over a wide area (generally at least 75 km radius from the radar installation) to be measured in near real-time. The information is usually colour-coded and displayed on a video screen, superimposed over a map of the catchment area and updated every 15 minutes. It can thus be used to obtain a good estimate of the areal distribution of rainfall, and forecasts of how this distribution will change can be made by extrapolating the movement and development of rainfall systems.

Weather radar has had a considerable impact on flood forecasting in England. Indeed, the benefits of flood warning proved to be main economic justification for the installation of the English network. Whether such a large benefit could be reaped in Scotland is less clear, however, bearing in mind the relatively low damage costs revealed by the above-mentioned survey. As has been indicated, the potential savings depend very much on the speed of response of the catchment, and the generally faster responding Scottish rivers do not offer as much scope for issuing warnings with a useful lead time.

The survey into flood damage costs collected information on the speed of response of the risk catchments and so the potential saving given a weather radar network can be estimated. Two estimates were made, one assuming the radar data are used purely for rainfall measurement, and one assuming rainfall forecasts are prepared using the radar data. The monetary benefits likely under these assumptions are shown mapped in Fig. 9.1 and 9.2 respectively, where it can be seen that most of the benefit accrues in the east of Scotland. The total annual benefit for Scotland was estimated to be £30,000 from the improved rainfall measurement, rising to £100,000 assuming a 4 hour rainfall forecast is made using the radar data (1983 values). This is considerably less than the equivalent values for England and Wales, namely £1·2 million and £3 million respectively.

In conclusion, it is clear that, although flooding is not a widespread problem in Scotland, there are localised areas where repeated flood damage occurs. Most of these are best served by the installation of efficient flood warning schemes and good meteorological data is essential if such schemes are to provide sufficient lead time for effective action to be taken.

ACKNOWLEDGEMENTS. The author would like to acknowledge the work of Mr C. Collier of the Met Office in collating and preparing the information on flood warning benefits in Scotland.

# The meteorological needs of the wind turbine industry

G. ELLIOT & S. M. BARTON

*National Wind Turbine Centre, National Engineering Laboratory, East Kilbride*

## INTRODUCTION

THE wind power industry requires accurate wind statistics in order that the resources available are fully utilised. Statistics are also required to optimise wind turbine design so that the industry can compete with other energy sources. In order to be cost effective, a wind turbine requires to be designed and built with a 20–30 year life expectancy. Thus, long-term weather statistics are essential to predict accurately both the energy output of the machine during its working life and its structural integrity. The most important variable is the wind speed since the power available is proportional to the cube of the wind speed. Accurate data are vital for meaningful evaluations to be made at a particular location. Wind direction is also important in site selection since site orientation and shielding effects caused by local topography can have a major effect on the wind speed.

## WIND DATA REQUIREMENTS FOR WIND TURBINE OPERATION

The power ($P$) incident upon a horizontal axis wind turbine (HAWT), per unit swept area of the rotor, depends on the density of the air ($\rho$), the wind velocity cubed ($U^3$), and is calculated as follows:

$$P = \tfrac{1}{2}\rho U^3,$$

where it is assumed that the turbine is operating in a steady wind speed $U$ ms$^{-1}$.

However, it is not possible for a turbine to extract all this potential power from the wind, as the air flow cannot be brought completely to rest by the motion of the rotor, due to conservation of momentum. The theoretical maximum power ($P_{max}$) available to a HAWT from the wind, known as the Betz limit, is:

$$P_{max} = (\tfrac{16}{27})(\tfrac{1}{2})\rho U^3,$$

or 59·3% of the power originally present in the moving air.

Turbine efficiency is represented by the power coefficient ($Cp$), and is defined as:

$$Cp = \frac{\text{power output from turbine}}{\text{power available in wind}}$$

The final expression for the power output of a given wind turbine in a constant wind speed $U$ ms$^{-1}$ is then:

$$P = Cp(\tfrac{1}{2})\rho U^3,$$

where $Cp$ (max) = $\tfrac{16}{27}$ or 59·3%.

This simple expression clearly depends critically on the wind speed. Thus, it is vital for the wind turbine industry to have a high level of accuracy in statistics used to determine projected power outputs. An error of 5 per cent in the wind speed becomes an error of 15 percent in the power estimation.

In order that the wind turbine designer can assess the forces and subsequent stresses occurring in a proposed wind turbine design, a knowledge of the maximum wind speed that will occur at a given location is necessary, together with an indication of how often this is likely to occur over the projected machine life time. The values that are presented radically affect the structural design and also the shut down criteria. Gustier sites are less desirable since not only is a wind turbine likely to produce reduced power due to its inability to respond to the fluctuations, but the resulting increase in the number of stress cycles reduces the life of the wind turbine due to fatigue.

The gust ratio, or the ratio of gust speed to mean wind speed, can be used to indicate effects on the wind due to local topography and terrain types. From the maximum gust in each hour an estimate of the maximum 100-year gust can be made. The survival requirements of a turbine at that site can then be established. It is not possible to record an absolute gust value due to the finite dynamic response of cup anemometers and so a gust time must be chosen within which the anemometer can record the wind speed. A three second gust duration has been adopted as being appropriate for wind energy calculations.

The wind speed near ground level is reduced by 'friction' effects of the land or sea over which it is passing. This is known as wind shear and the wind speed generally increases with height. Details of wind shear are important for predicting wind speeds at the centre of a wind turbine rotor (hub height), from the height at which data were measured (Hunter & Collins, 1987). In general hub and anemometer heights will not coincide, since the standard meteorological measurement height is set at 10 m. It is clearly the former which is of interest in power performance calculations.

The variations of wind speed with height can be severely affected by topography, therefore the means of extrapolation is not at all clear. In order to measure wind shear, a series of anemometers at various heights above the proposed site are required. A 10 per cent difference between anemometer and hub height wind speeds will give a 30 per cent difference in calculated power output between the two heights. The largest machines currently in operation have rotor diameters of 60 m and significant wind force variation is experienced on the rotor blades between the highest and lowest parts of their rotational sweep. Wind speed is measured at the highest, lowest and mid points of such rotor cycles.

An investigation into the long-term meteorological statistics shows variations of $2\ \text{ms}^{-1}$ in mean annual wind speed at several sites over a 20-year period. This cycle will occur within the working life of a wind turbine. Knowledge of such trends is necessary since they will have a significant effect on the power potential of any proposed site (Palutikof et al., 1986). Details of wind direction are important when site orientation is being planned. This is particularly true if the site is to support more than one turbine, or if there are obstructions in the vicinity, as shielding effects must be considered in order to extract the optimum amount of energy from the site.

G. Elliot and S. M. Barton

# THE SUITABILITY OF WIND DATA AVAILABLE FROM THE MET OFFICE

The Met Office currently has some 120 anemometer sites throughout the UK, of which around 40 are located in Scotland. Most of these are situated at airports, ports or centres of population and mostly at low lying ground around our coasts or in valleys. Only three stations lie between 100 m and 200 m, while only five are 200 m or more above sea level, a distribution which reflects the bias towards low elevations of all meteorological observations in Scotland (Fig. 10.1). These stations provide

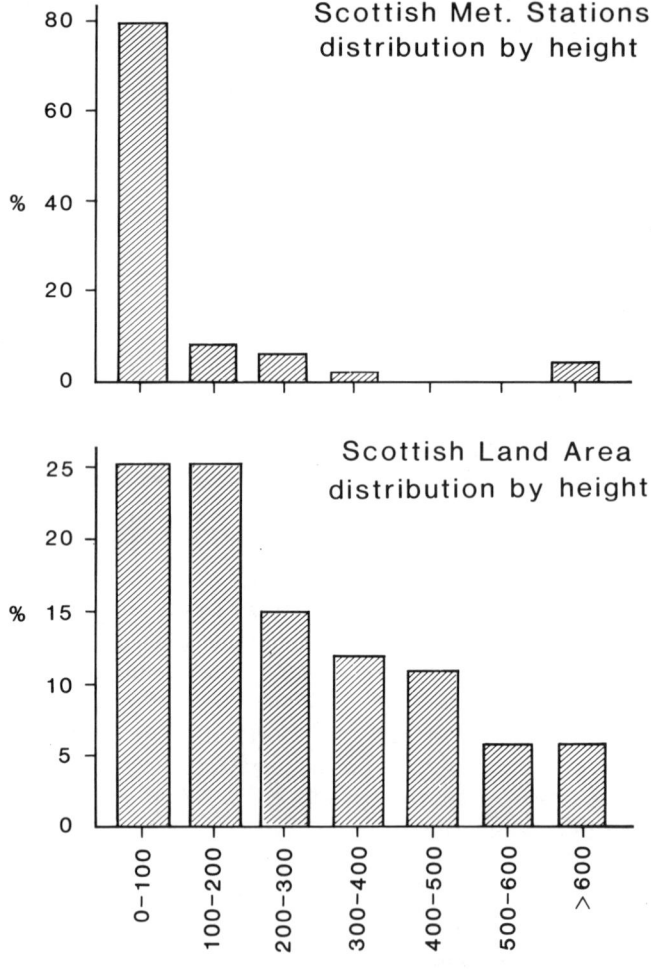

FIG. 10.1. Distribution of land area and meteorological stations in Scotland showing low elevation bias.

mean hourly values of wind speed and direction, as well as speed and direction of the strongest gust each hour. Observations of the wind climate at upper air stations are recorded four times daily, by means of balloons supporting instrumentation which ascend to a height of 900 m.

Current Met Office wind data is averaged over one hour. This presents problems for wind power developers as the mean hourly wind speed is inadequate in itself to provide an accurate estimate of average wind power. The variability within the hour, such as the standard deviation is also required. In the field of wind energy the method of collecting data has been resolved for some years by the 24 countries who collaborate under the auspices of the International Energy Agency. As a result a recommended practice, which standardises on 10-minute averaged data and a method of bins for sampling, has been produced. This is now almost universally used by the wind energy community.

Hourly averaging presents problems when data correlation between available meteorological data and that measured at turbine locations is required. A considerable amount of data manipulation may be required to arrive at meaningful hourly data from recorded 10-minute statistics in order to make comparisons, which can be costly in both manpower and computing time. When using low cost computer based data acquisition equipment to collect and derive 10-minute statistics, typically only some 60%–70% of the available data can be collected. Averaging over a longer period would increase the quantity of data actually recorded, but would also demand greater computer storage capacity prior to processing, with a resulting increase in data acquisition costs.

Diagrams which map the average wind speed around the British Isles have been presented in various documents, particularly for estimating wind loadings on buildings. These diagrams reduce the already flimsy data to sea level in order to plot the contours on a common datum. As a result, quite erroneous analysis has been conducted in estimating the size of the available wind resource as a measure of the market for wind turbines. The latest of these has recently been published by the EEC using isovent data to estimate the market in the ten member countries. The analysis misses the fact that the data is corrected to sea level and under-estimates the size of the resource so much as to make the study meaningless. Fig. 10.2 shows the isovent map for Scotland with a few additional data points. The differences between the values are extremely significant. It must be borne in mind that the wind power is proportional to the cube of the wind speed. This makes the actual values of power available much greater than those derivable from the isovent diagram. For example the National Wind Turbine Centre site at Myres Hill lies on the 4·5 ms$^{-1}$ isovent whilst in practice the annual mean is 7·9 ms$^{-1}$.

Only about half of the anemograph stations in Scotland have been currently updated and equipped with Digital Anemograph Logging Equipment (DALE) Fig. 10.3 shows these locations which are ideal for power production estimates. The remaining stations produce pen charts which require manual evaluation in order to establish mean wind speeds and directions. The quality and accuracy of manually analysed wind statistics produced from anemograph stations depends largely on

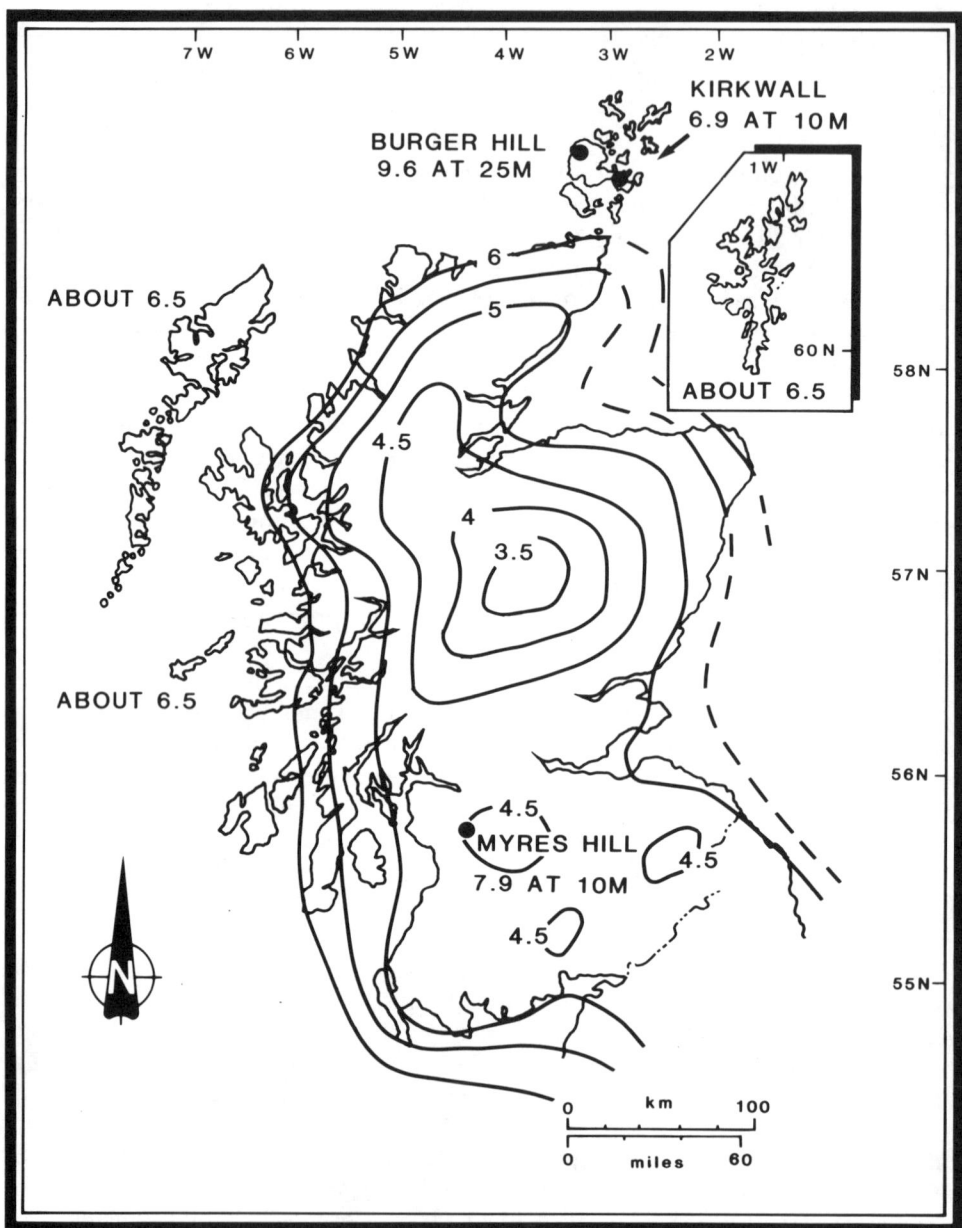

Fig. 10.2. Isovent diagram for Scotland.

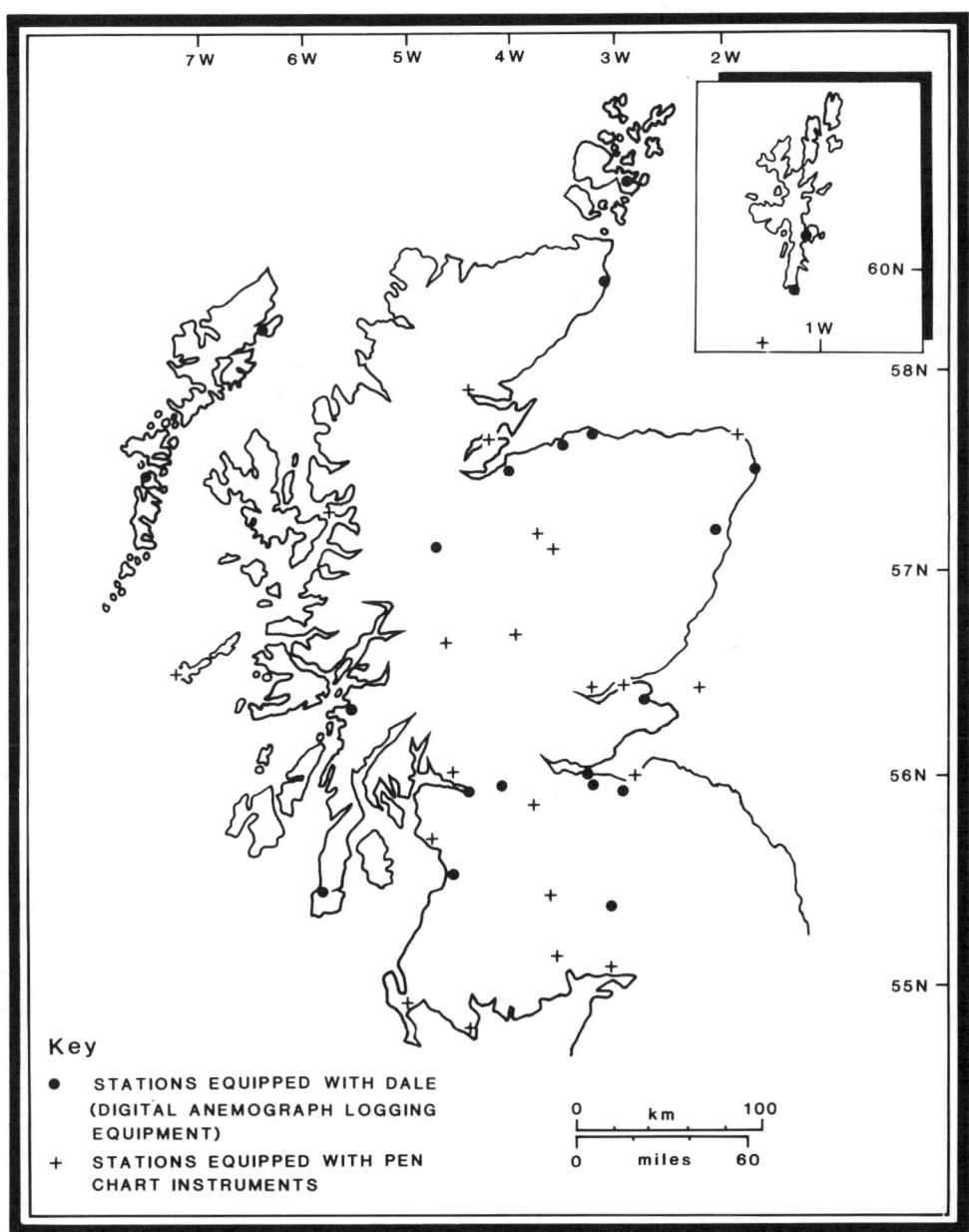

FIG. 10.3. Distribution of anemograph stations in Scotland.

individual interpretation of the recorded traces and hence are subject to inconsistencies and error. Wind power statistics obtained from such charts could possibly be in error by 100 per cent or more and the wind turbine industry, therefore, gives preference to digitised values of wind speed and direction available from DALE stations.

## INADEQUACIES OF AVAILABLE DATA

The windiest areas in Scotland generally occur on the coast or at the higher elevations inland and it is obviously these sites which are of the most interest to the wind turbine industry. However, only 8 of the 41 anemometer stations in Scotland lie above 100 m. This leaves large areas of high level terrain, which are potentially suitable for wind energy development, for which no statistics are available. The exclusive use of DALE statistics further restricts the land area covered by meteorological data, conflicting with the desires of wind power developers to see the network extended to cover all of the UK. In particular the north and west coasts of the Scottish mainland are devoid of recorded meteorological data in what almost certainly, due to the prevailing climate, are the windiest locations in Britain and possibly in Europe.

Various attempts to transpose wind data from the upper atmosphere have been attempted, including the European Wind Atlas (Petersen & Troen, 1988). However, in complex terrain, such as in Scotland, it is unlikely that meaningful information will be provided. Transferring data from one site to another has been attempted and complex computer models have been written with this in mind. Local effects, such as wind-channelling, due to surrounding rough terrain and shielding due to obstacles will influence wind speed and direction recorded at the anemometer site. Corrections to the original data will be necessary in order to achieve best results on the proposed site and the accuracy of the resulting data becomes questionable.

Wind speed data are widely measured using cup anemometers, which must operate accurately in a wide range of climatic conditions. A phenomenon which has been observed and reported by a number of researchers in the wind energy field is the apparent drop in power output from some wind turbines during periods of rain. Since power output is always measured as a function of wind speed, the effect is either a genuine alteration of the aerodynamic performance of the wind turbine or the anemometers are giving an enhanced reading. The effect is quite significant and apparent power reductions of as much as 30 per cent have been reported. If this proves to be in part an anemometer phenomenon, then the data that has been recorded from meteorological stations will be further treated with suspicion. Other effects, such as icing, may cause anemometers not to work or to read incorrectly; however these incidents are never identified. Wind energy researchers have also reported significant changes in the calibration of anemometers due to wear in the bearings.

It is, therefore, important that these instruments are regularly calibrated, at six-monthly intervals, to ensure that the data recorded is meaningful. The Met

Office, however, carry out a three-year programme of instrument calibration. It is reasonable to expect wear of the bearings to occur within this calibration period with a corresponding effect on the data recorded. The average lifetime of anemometer bearings is 2–3 years, depending on the model involved. Re-calibration at other times is only undertaken by the Met Office if data produced by an anemometer is called into question. Anemometers at present are also only calibrated up to wind speeds of 25–30 ms$^{-1}$ and it is believed that some designs are incapable of reaching the very high gust speeds that are experienced from time to time. The design survival wind speed for wind turbines is 60 ms$^{-1}$. The highest gust speeds recorded must be treated with suspicion if they exceed the maximum calibration of the anemometer.

Research into the accuracy and dynamics of anemometers is being carried out at the UK National Wind Turbine Centre and a free jet calibration facility has been developed in order that all anemometers used on turbine sites can be regularly calibrated in a cost effective manner against a national standard. Work is also being done to establish a standardised calibration procedure for all National Wind Turbine Test Stations within the European Community.

## CONCLUSIONS

This paper has attempted to highlight the meteorological needs of the wind energy community and to show how these needs differ from the standard data produced by the Met Office for weather forecasting purposes. As the ultimate aim of the wind turbine industry is to predict long term power availability, the emphasis on wind data is naturally much greater than that of the meteorological community. The data are required in great detail in order to carry out sophisticated analyses necessary to determine the long term economics of this new and emerging industry. The geographical anomalies between existing meteorological data stations and the high wind speed, high ground areas, where wind energy generation might be pursued is quite marked and an extended network of measurement stations is highly desirable.

The future market for wind turbines is at present estimated to be extremely significant worldwide, and the desire is for British manufacturered products to be exported to fill these markets (Jaras 1987). However, the basis of a sound export business is a sound home market and good data are necessary to establish this. Scotland is likely to be the main centre for this home market due to its high wind speed climate. There are also numerous tracts of low quality land on which wind energy schemes could be developed, to the enrichment of the nation as a whole, by creating new employment and new wealth, as well as providing a valuable source of pollution-free energy.

# SECTION FOUR

# INDUSTRIAL APPLICATIONS

# Sources of United Kingdom weather data

F. SINGLETON

*Meteorological Office, Bracknell*

## INTRODUCTION

In common with other countries the United Kingdom Met Office receives weather data from a variety of sources in different ways and with various degrees of timeliness. In the virtually instantaneous category of data reception there are radar and satellite systems. The pictorial information is very valuable to forecasters and, for some purposes, the quantitative data have industrial applications. Radar derived rainfall data have been archived since 1984. For many purposes, however, there is still a need for the conventional meteorological, instrumental and descriptive, observations of, for example, temperatures, winds, rainfall, weather, visibility and cloud.

## THE SYNOPTIC NETWORK

For day-to-day forecasting purposes data are received, quality controlled in a rather limited fashion to meet operational needs, and re-distributed in very near real-time. The ability of the communication system, even in its modern computerised form, to handle large amounts of data is limited. Although some parameters (such as temperature) can be measured and input automatically to the system, there is expected to be a continuing need for human input to describe the weather, including cloud type and visibility, so limiting the number of locations from which such observations are available. The ability of the human-being to assimilate data is likewise limited, although computer models are becoming increasingly important as aids to the forecaster in decision-making (Golding, this volume). However, in the final analysis there is still a need for human assessment and judgement. Consequently, the synoptic network (Fig. 11.1) has many gaps and, while it meets many needs for much of the time, it cannot meet all needs for all of the time.

In practice, the synoptic network has a sufficient density of data to permit the definition of most weather features of significance that cross the United Kingdom at all hours of day and night. Increased observing schedules and, therefore, data volumes during the daytime allow for rather better definition of weather at times likely to be of interest and all data needs will not be met by the synoptic observing network of discrete observations at specified locations. Some gap filling is possible by means of radar and satellite techniques and, for some particularly specialist purposes, extra data may have to be acquired or made available by the main beneficiary. For example, in order to aid the forecasting of road conditions, the monitoring of road surface temperatures and of visibility along roads could be

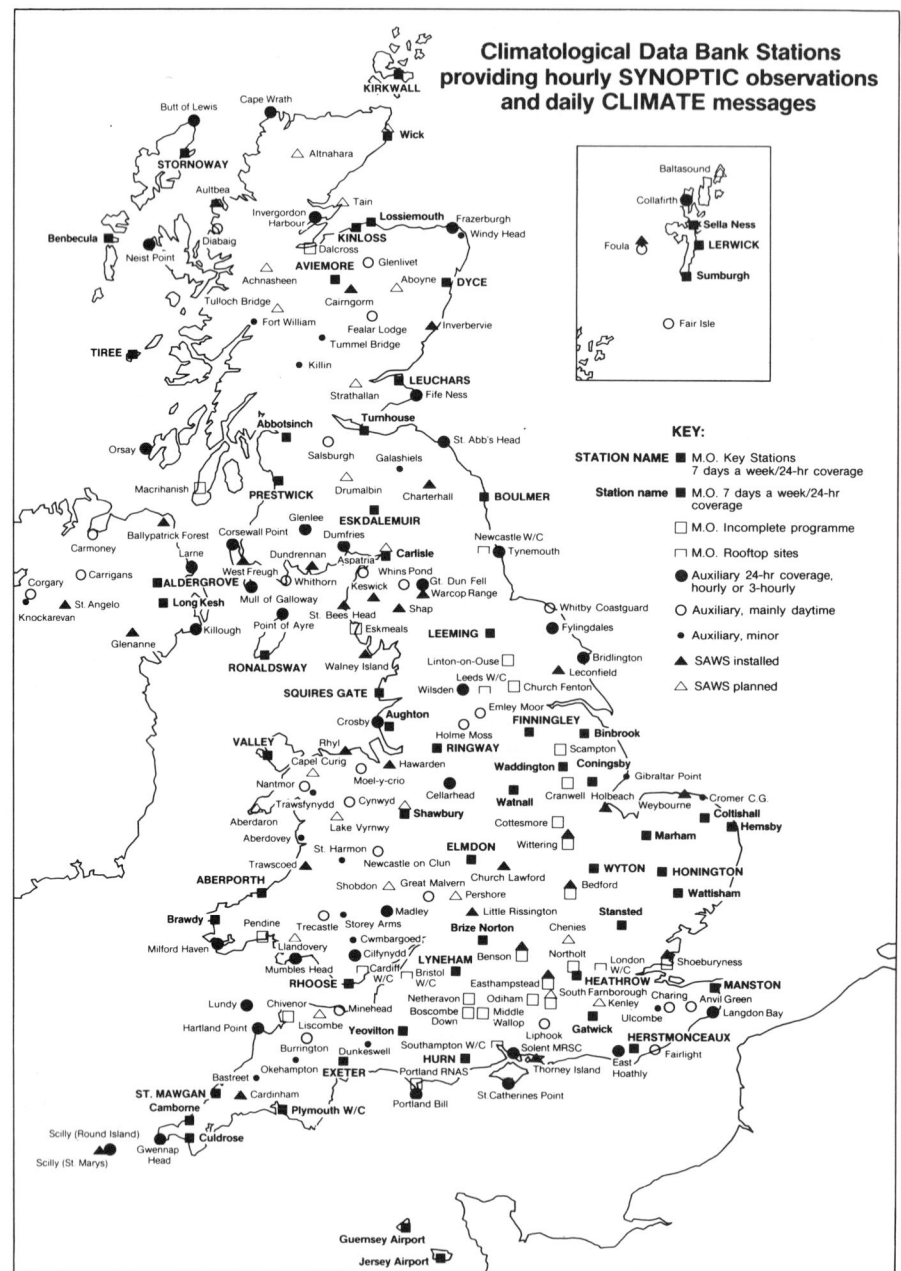

Fig. 11.1. Distribution of synoptic stations providing climatological data (April 1987).

undertaken, not unreasonably, at the expense of those responsible for the road system and who need the services for which such data can be an important input.

To the forecaster, and for many operational applications, weather data are highly perishable in that the focus is upon what is happening now and what will happen in the foreseeable future from a few hours to a few days ahead. However, many applications of weather information require a knowledge of what has actually happened at a given location at a given time, or information on statistics of weather, extreme values and probability of occurrence of various events. To this end the Met Office archives, both on hard copy and on computer, all available United Kingdom synoptic data. This archive, like the network itself, is a somewhat variable, and difficult to define, entity. Thus, on an hourly basis, there are on magnetic tape data for 6 United Kingdom stations back to 1949, hourly data for 38 stations back to 1957 and, currently, data are archived from 67 stations. A further 100 or so stations have data which either do not cover the 7-day week or the 24-hour day. All these synoptic data are collected in real-time from the synoptic messages passed over the telecommunications network. Changes in requirement, particularly by the RAF, automation of light vessels, and general cuts in Civil Service manpower all result in officially manned stations opening and closing from time to time. The network is, therefore, somewhat transient in nature.

## THE CLIMATOLOGICAL NETWORK

For archival purposes and subsequent use, the synoptic network is enhanced by data from the co-operating climatological network to give a total of some 600 stations reporting daily values of extreme temperatures, rainfall totals, and sunshine totals together with information on ground level temperatures, soil temperatures, state of ground, whether snow has fallen and so on (Fig. 11.2). Data kept in the climatological archives, whether originating from synoptic observers or voluntary climatological observers, are all subjected to rigorous quality control procedures for self consistency and areal consistency.

Some 30 stations have such daily data on the computer archive back to 1931 or earlier, while a further 300 stations have data on the archive for at least 25 years. Because the Office has even less control over this network than for the synoptic network, the lifetime of the individual stations is very variable and there are, in fact, data from over 900 different sites on the climatological archive data sets.

The climatological network has its origins firmly in the last century when the Scottish Meteorological Society, in particular, did much to develop an observing network, observing standards, observational techniques and validation or quality control of the data. Because of the complexity of the network, the many different contributors, and their different degrees of technological awareness, the method of data collection has been kept simple and is still via the traditional tabulated forms nowadays keyed to computer by the Met Office. Some of the voluntary observers are already using automatic weather stations and could provide data in a computer medium, while others are happy to key data to a local PC. However, it will be some

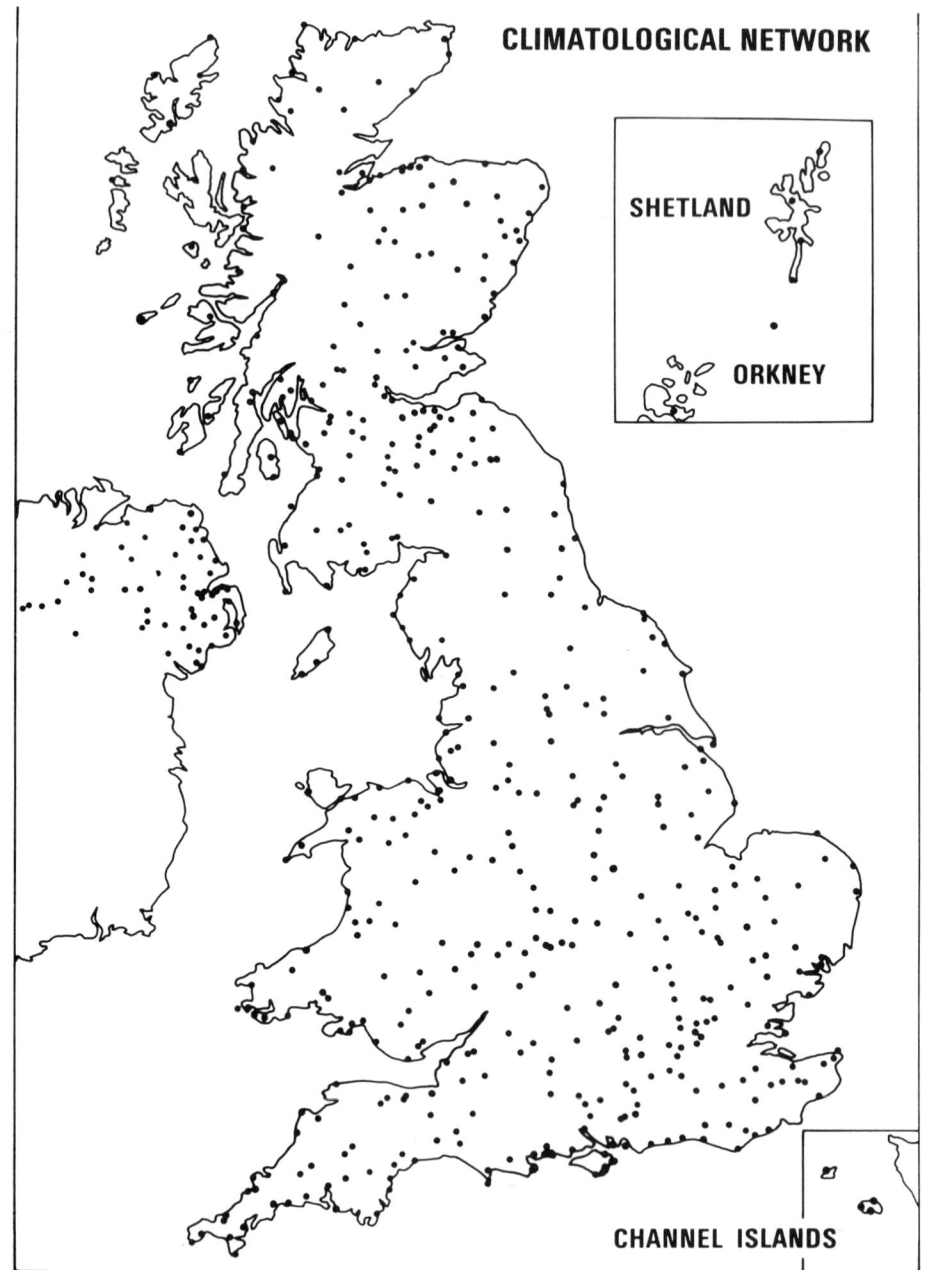

Fig. 11.2. Distribution of climatological stations in the UK (April 1987).

while before climate data collection becomes as automated as the collection of synoptic data.

## RAINGAUGE NETWORK

For rainfall the climatological network is further enhanced by a raingauge network largely manned by co-operating observers. The 5000 strong rainfall network, like its general climatological counterpart, owes much to our Victorian fore-fathers and, in particular, G. J. Symons who formed the British Rainfall Organisation in 1860 creating a network that, by the turn of the century, numbered some thousands of gauges. As with the climatological network, rainfall data are still collected on a monthly basis by means of postcards sent either to Water Authorities, River Boards or, in a minority of cases, direct to the Met Office. These data are all keyed to computer. Again, the transition to faster, and less labour intensive, methods of data collection is probably some way off.

The Met Office receives data from a relatively small number of continuously recording raingauges, most of the data being from tipping bucket raingauges where the data are recorded onto a solid state recording medium but some are still in paper form from the traditional Dines tilting syphon raingauge. There are more raingauges measuring rainfall totals over short periods, many of these being run by those concerned with the management of water resources. There is probably a need for the Met Office to get together with the various bodies concerned to acquire some, at least, of these data which are stored by some authorities but only used in an operational sense by others and then lost. However, there are many problems of compatibility between the various computing systems and recording media. The long-term answer might well be to use radar as a means of deriving short-period rainfall data in conjunction with the dense daily gauge network. Given the existence of radar for other purposes, this could prove to be the most cost-effective way forward.

Like the other networks the raingauge network varies very greatly from time to time, the Met Office having little or no control over the opening and closing of most of the stations. The bulk of the computer-based rainfall archive dates back to 1961 and contains a total of some 14,000 stations. Of these, however, only 1560 have continuous records back to that year. The current number being archived is just over 5000. A total of 182 long period stations have had their data keyed back to 1925 at least.

## OTHER NETWORKS

The anemograph network comprises some 160 stations with rather more than half being owned by the Met Office. Like the recording raingauge network there is a mixture of conventional chart recorders of wind speed and direction with a number of electronic logging systems. For climatological purposes a decision was made many years ago to use anemographs to extract hourly mean wind and maximum gust data.

Most UK wind statistics relate to these parameters and most design criteria are based upon them. Information about winds over shorter periods than an hour or gusts of different duration from the standard 3 second gusts are usually obtained by means of empirical relationships with the hourly mean. The digital wind logging equipment does, however, allow the recording of minute by minute wind data and these have been archived since 1983. However, the problem here is simply one of storage and data management and it has never been intended that these data should be kept in perpetuity. Current considerations of needs for wind data are confirming that the prime archival needs and industry requirements would be met by means of hourly means and gust data. In the future one-minute data may only need to be recorded at a limited number of stations—in the region of 20–30 for the whole of the United Kingdom.

Sunshine data are recorded at some 340 stations using the Campbell–Stokes recorder to give sunshine duration. Although these data are highly regarded for their publicity value by holiday resorts, and are useful in maintaining historical data series, they are of no great value in any industrial or scientific sense where the observations really required are of radiation. Although some radiation data can be deduced or implied from sunshine totals, it would clearly be preferable to have a comprehensive network of radiation instruments. Currently, solar radiation is only measured at around 38 stations, of which rather less than half are at Met Office sites. Of these 38, about half report hourly radiation data and the remainder give daily values. All the Met Office sites measure global and diffuse radiation and all available radiation data are held on the computer archive. In the fulness of time it is hoped to have a comprehensive radiation network with the Campbell–Stokes recorder not being used in any truly scientific sense.

## MET OFFICE DATA ARCHIVES

There are various sets of derived data and software exists within the Met Office to undertake analyses of various forms using the climatological archives. An important new facility is a climatological data set of evaporation parameters. These are derived from the Met Office rainfall and evaporation calculation system known as MORECS. The MORECS program provides areal values over $40 \times 40$ km squares of potential and actual evaporation, soil moisture deficit for different soils and crops, effective rainfall and sunshine.

The Met Office also has a number of worldwide data sets some of which are obtained simply by storing all data received for synoptic purposes on the global telecommunication circuits and some of which have been obtained from other countries, mainly the USA. There is an extremely comprehensive marine archive dating back to the 1850s, all transcribed to computer and now numbering around 70 million observations, allowing analyses to be undertaken for oceanic areas worldwide. There are, for example, archives of sea surface currents, sea surface temperatures and upper air archives.

In summary, the computer based archive of synoptic data dates back to the late

1950's generally and to 1949 for a small selection of stations. Currently there is, on average, one station per 1500 km$^2$. Data are archived from anemograph and recording raingauge networks of similar density. Daily climatological data are archived for about one station to every 400 km$^2$ on average, with most of the archive dating from the late 1950's. Data for 30 stations are on computer file since at least 1930. Daily rainfall data are on computer files at a density of one station per 50 km$^2$ since 1960. A selection of stations date back to 1925 or earlier. Radiation data are available for one station per 6000 km$^2$ but sunshine data from one station per 700 km$^2$. Radar derived rainfall data have been archived since 1984 from the far from complete radar network. In addition, Met Office document archives in Bracknell, Edinburgh and Belfast retain most manuscript returns of meteorological data under the terms of the Public Records Acts.

# The application of Scottish weather records

R. C. TABONY & P. A. D. BROWN

*Meteorological Office, Edinburgh*

## INTRODUCTION

MOST enquiries relating to past weather in Scotland are handled by the Met Office in Edinburgh which holds all the official weather records made in Scotland. The number of climatological enquiries now amounts to 5000 per year and has grown steadily over the years (Fig. 12.1). This is a trend which the Met Office is keen to maintain as it indicates a growing awareness of the benefits of past weather data to the nation. The sections of the community which make the most use of climatology are legal and insurance, building and contruction, and education, as shown in Fig. 12.2(a). In viewing these figures it needs to be borne in mind that there are also a substantial number of agricultural and marine enquiries, but most of these are dealt with by specialised bureaux in Bracknell (see Callander this volume). A breakdown of climatological enquiries by revenue (Fig. 12.2b) shows that, while the legal and building categories maintain their ascendancy, the educational queries are mainly related to project material for Higher level geography courses and yield little income. Enquiries linked to design and planning are relatively few in number, but this is an area in which climatology can make a significant contribution to the national economy. In the remainder of the paper, the services available to the three main types of enquiry—legal, building and planning—are described in more detail and are followed by a discussion of some of the consultancies on offer.

## LEGAL AND INSURANCE

Much of the business of the legal and insurance professions is related to accidents necessitating a statement of the weather prevailing at the scene of an incident. As many accidents are caused by ice or snow, conditions on the ground are often as important as those overhead and a knowledge of antecedent conditions is required in order to assess whether there was any negligence in clearance operations. Many insurance claims are also related to structural damage in which wind is the primary agent.

The requirement to know the weather conditions at a particular point and particular time places great demands on the climatological network. Singleton (this volume) indicates that there will often be a daily station within a few kilometres which records the rainfall and maximum and minimum temperatures over a 24 hour period, together with a brief weather diary and a full observation at 09·00 GMT. This information is used in conjunction with hourly observations from the nearest Met Office station to make deductions about conditions at the required time and

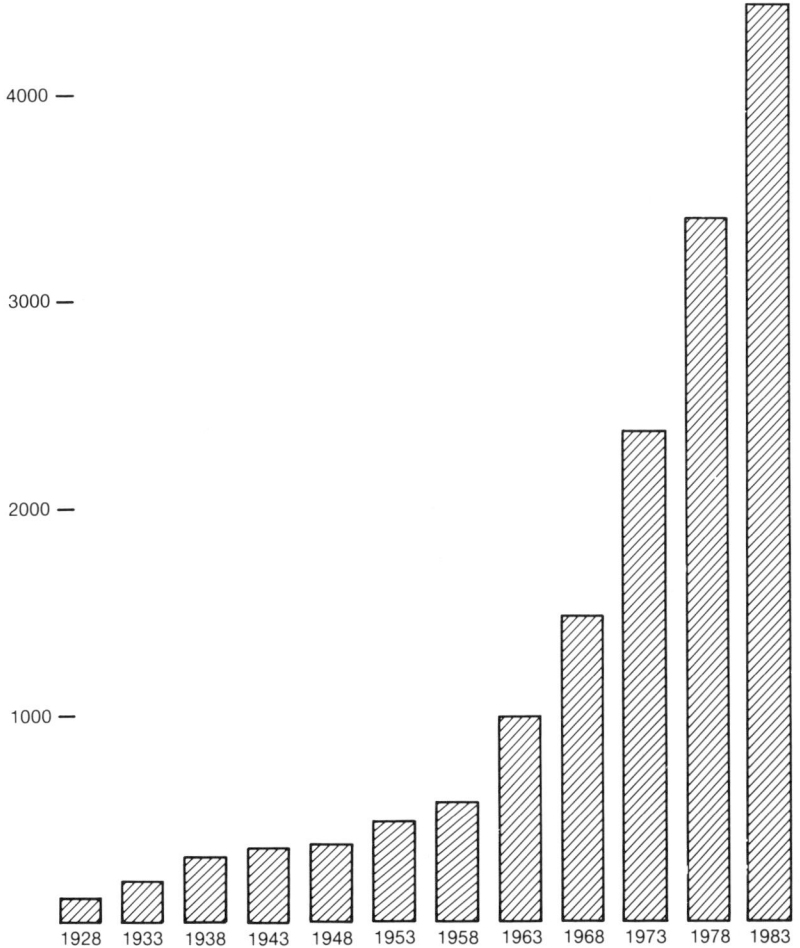

FIG. 12.1. Annual number of climatological enquiries dealt with at the Edinburgh Met Office.

place. The Met Office site may be within a reasonable distance in lowland areas, to which most enquiries relate, but it could be unacceptably distant or unrepresentative in some upland districts. Satellite images may be helpful in resolving difficulties on some occasions.

Improvements in estimates of 'point' weather will be taking place in the near future. First, the installation of automatic weather stations will result in a large increase in the amount of hourly data available. Secondly, the introduction of weather radar to Scotland will provide comprehensive coverage of rainfall in both space and time. Thirdly, the installation of surface sensors in connection with the 'Open Road' forecast scheme will supply better information on ground conditions. The model used to forecast road temperatures is also being linked to the

(a) PERCENTAGE BY NUMBER

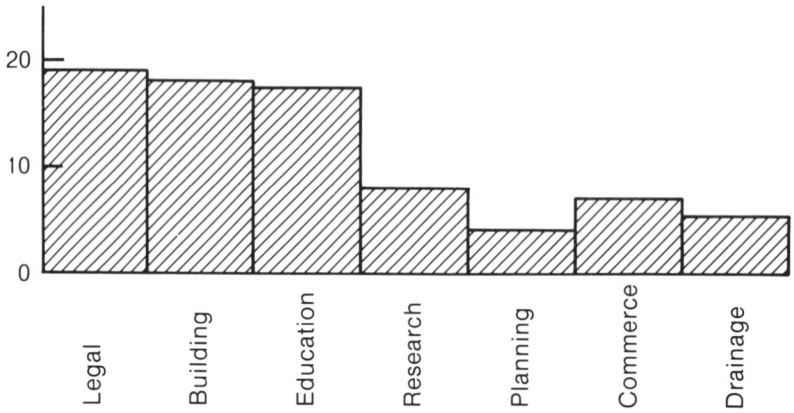

(b) PERCENTAGE BY REVENUE

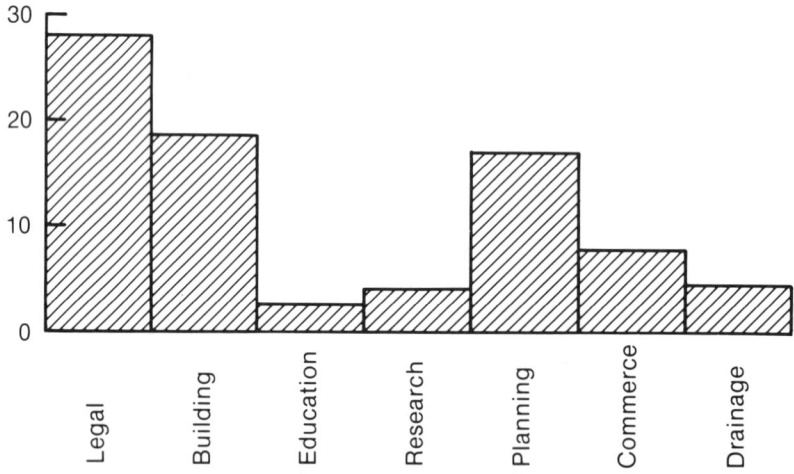

FIG. 12.2. Types of climatological enquiry at Edinburgh by number and by revenue.

climatological data sets to provide extra support in this area. It needs to be pointed out, however, that the reliability and range of parameters recorded at automatic weather stations do not match those provided by manned stations. Reports of phenomena such as state of sky, lightning, whirlwinds, precipitation type, depth and drifting of snow, for example, will not be available from automatic weather stations.

In spite of the improvements mentioned above, it is inevitable that the precise conditions at the scene of an accident will remain a matter of opinion. The Met Office will always offer opinions in the covering letter to a certified statement and is prepared to back these up through court appearances. Insurance claims related to the weather occur on a semi-regular basis and the regular monitoring of relevant weather parameters will save the insurance industry many time-consuming telephone calls. To this end the Met Office is developing a publication intended for routine despatch to insurance companies and it is hoped that they will support this new venture.

## BUILDING AND CONSTRUCTION

There are four main stages in any major construction project, namely design, planning, operational and post-mortem, the latter including an assessment of the performance of the completed structure. Various aspects of the tendering/planning and operational phases are discussed by Prior and Cumming (this volume). It is not only forecasts which are of value in the operational stage, as the monitoring of the weather is an essential aid to the management of a project and any post mortem enquiries. The Met Office has developed a monthly booklet known as '*Metbuild*' for just this purpose and it contains the number of hours in the working day when various weather parameters have exceeded certain industry-related thresholds. It is available only for those stations where hourly observations are available, but the expansion of this class of station via automatic recorders should widen its scope. Although *Metbuild* will tell a reader whether adverse weather has been more or less frequent than usual, it does not contain sufficient information as always to permit a settlement of extension of contract claims. The Met Office can be approached direct to help in the adjudication of these matters.

A possible improvement in the management of projects would be the installation of an automatic weather station on site. This would remove any ambiguity which might arise from the use of data from the nearest official station. Advice would still be needed to assess the average climate of the site and hence decide whether the conditions experienced were exceptionally adverse or not.

## DESIGN AND PLANNING

Many enquirers realise that the Met Office can provide raw data but are unaware that a wide variety of analyses is also available. By making use of these at an early stage much money can be saved by first considering the feasibility of a project and then avoiding the possible disaster of under-design or the unnecessary expense of undue caution.

The first restriction that can be removed is the need to deal only with directly observed elements. Derived parameters such as degree days, wind chill or driving rain can be provided. One of the simplest forms of analysis is a frequency distribution of a single element. The distribution of dew points, for example, can be used to provide information on the overall frequency with which humidifiers need to be used in air conditioning systems. Multi-dimensional frequency distributions can be even more useful. For example, temperature v humidity by hour and month can indicate the daily and seasonal changes in the mean demand on central heating systems, together with the likely variability. The probabilities of rare events can be obtained from extreme value analyses which estimate the values of elements likely to be exceeded once in so many years. Such return period analyses of wind, temperature and rainfall provide design criteria for a wide range of engineering activities. The 2-minute rainfall to be expected once every 5 years, for example, is commonly used to decide upon the number of drains to install in a paved area. The Building Research Establishment has also written a computer program known as Strongblow which enables topographic effects to be taken into account in the estimation of design wind speeds (Cook et al., 1985).

Another useful form of analysis involves the selection of occasions when one or more elements exceed various thresholds. This is the basis of the *Metbuild* downtime summaries, but more sophisticated analyses are possible. Guidance on the icing of overhead transmission lines, for example, can be obtained by selecting occasions when temperature, humidity, wind speed, visibility and precipitation fall between certain limits. High humidities combined with sudden rises in temperature can be used to identify occasions when condensation is likely while the occurrence of low temperatures, high humidity or rain indicate conditions unsuitable for exterior decoration. A further sophistication takes the form of spells analysis, and this is commonly used to find the probability of experiencing weather windows suitable for the execution of sensitive engineering activities.

The most elaborate type of analysis is perhaps the provision of a frequency distribution of occasions satisfying various selection criteria. A recent example involves Grampian Regional Council who sought advice on the quantity of salt required to keep roads clear of snow and ice in the worst winter in five. 30 years of weather records were used to infer road conditions and hence the dose of salt required on every winter's morning. These figures were used to obtain seasonal totals of salt usage which were ranked to provide the median and 'once in 5 year' estimate of salt demand.

## CONSULTANCIES

The Met Office can offer an impressive range of sophisticated analyses, but a disadvantage is that they are only available for locations with a long run of data. It may be that advice is needed for a site where the application of records from any of the permanent stations is clearly inappropriate. A solution might be to install an automatic weather station for a trial period to obtain a limited amount of data. The Met Office would then be able to make comparisons with observations from established stations and deduce a long-period climatology for the site in question. A good example of this occurred in connection with the possible ski development at Aonach Mor near Fort William, where a knowledge of the likely winds was required to establish the feasibility of a proposed gondola lift system. An anemometer was installed at the location of the top station and comparisons were made with the wind in the free atmosphere derived from (radio-sonde) balloons released from Stornoway. This enabled the long-period wind regime at Aonach Mor to be deduced.

A relatively new venture for the Met Office is the use of regression analysis to convert weather data into forms that are of more direct use to the customer. There are few activities in life which are completely independent of the weather but in many cases they are masked by other factors. A statistical regression analysis is capable of uncovering the weather dependence, however, and this technique offers possibilities in the world of commerce, especially the retail trade. Once the particular aspects of weather which affect a business have been identified and quantified, the weather dependence of sales figures can be removed and the true performance of a business monitored. The statistical relations can then be used to convert forecasts of weather into predictions of sales. This offers many advantages in the form of reduced storage and transport costs, better stocked shelves, premium pricing opportunities and tactical advertising. It will be readily appreciated that this technique offers enormous scope for many sections of the economy to profit from the application of weather records.

## SUMMARY

The demand for past weather is increasing steadily with time as more and more people become aware of the potential benefits of climatology. The majority of enquiries belong to the legal or construction categories which require a knowledge of site specific weather and the advent of automatic recorders will result in an improved service for this type of query. A wide range of analyses are available to help in the design stage of many projects and this is an area where the application of climatology has a very favourable cost:benefit ratio. A recent venture is the use of statistical regression to uncover the weather dependence of sales figures and this technique offers the prospect of extending the benefits of climatology to ever widening sections of the economy.

# The effects of adverse weather on the construction industry

## J. R. T. CARSON

*Miller Construction Ltd.*

### ADVERSE WEATHER AND CONTRACTS

THERE is no precise definition of exceptional adverse weather to be found within the construction industry. Abrahamson (1979), a leading authority on construction law, defines it thus: "conditions may be exceptional either in their intensity, for example a storm which disrupts the contractor's work, or in their duration, for example an abnormal number of rainy days in the place and season of the contract".

In order to define adverse weather it is necessary to examine the three main forms of contract generally used in the UK. These are:
  (1) *Institution of Civil Engineers* (5th edition). For civil engineering works (June 1973) (Revised 1979), commonly known as ICE (Fifth).
  (2) *Joint Contracts Tribunal for the Standard Form of Building Contract* (1980 Edition), commonly known as JCT 80.
  (3) *The General Conditions of Government Contracts for Building and Civil Engineering Works* (2nd Edition 1977), commonly known as GC/Works/1.

The contract entered into by the client and the contractor to execute a construction project includes not only the various aforementioned standard forms but also drawings, specifications, site investigation data, method of measurement, special requirements and bills of quantities.

The terms and conditions of the standard forms require the contractor to take the risk involved and make due allowance for adverse weather in his tender. Each contract allows the contractor additional time to complete the works if events outwith his control thwart progress. This includes exceptional adverse weather. As more and more contracts are implanted with onerous liquidated damages clauses, and with clients being more willing to inflict these charges, contractors must protect themselves against such events. Although exceptionally inclement weather will not allow the contractor any recovery under the three terms of contract, it does allow the time for completion to be extended thus delaying the date on which damages fall due.

The definition and nature of adverse weather appears to vary under each of the aforementioned terms of contract. These are:
  ICE (Fifth)    "exceptional adverse weather conditions"
  JCT 80         "exceptional adverse weather conditions"
  GC/Works/1 "weather conditions which make continuance of
                  work impracticable"

The difficulty in arriving at a precise definition of exceptional adverse weather is further complicated by the persons appointed by the client to administer the contract, who each have their own interpretation. In ICE (Fifth) the engineer who supervises the contract for the client, who is usually also the designer of the project, defines conditions, while under JCT 80 it is the supervising architect, and under GC/Works/1 the supervising officer, who effectively manages the contract. Each one of these will have their own definition of adverse weather. From the point of view of the engineer, architect, and supervising officer, the skilful use of the adverse weather clause in the contract, by encouraging and awarding additional time to contractors for weather for which there is no direct entitlement to cost reimbursement, may mitigate additional costs to their clients. On the other hand, the contractor would rather claim under other clauses of the contract which would entitle him not only to time extension but also costs.

## ALLOWANCES FOR ADVERSE WEATHER

It follows that the contractor should take due allowance in his programme and price for the effects of expected inclement weather, and its timing, on the various construction operations. The three forms of contract previously mentioned cover a wide variety of construction work. Civil engineering has different requirements regarding allowances, effects, and timing to that of building constructions since most, if not all, of the former's activities are subject to weather, whereas those of the latter are affected only in part.

In civil engineering contracts the contractor would normally programme the weather-susceptible operations for the better parts of the year. Activities such as muckshifting in weather-susceptible soils, road formation work, blacktop activities, working in and around flood-prone rivers, and concreting are all ideally programmed from April to October. Indeed, it would not be too extreme to say that contractors would not work during December, January and February given the choice, but constraints of time for completion compel them to do so. Civil engineering contractors will normally make allowances for inclement weather by reducing the output for winter working for both machines and labour with corresponding increases in wastage of materials, as shown in a sample Working Week Sheet (Fig. 13.1). The natural conclusion from this is, for example, if an eighteen month contract was let in October with two intervening winters this would result in a more expensive contract than the same contract let in April. The extent to which a contractor will reduce his outlay is obviously dictated by the location of the contract. The downtime expected on the west coast of Scotland will be considerably more than the time lost in, for example, Dunbar which is in one of the driest areas on the east coast. In building construction contracts the contractor will determine the portion of the contract subject to weather and make due allowances, the wind and watertight stage of the contract being the critical phase for non-weather susceptible operations.

## MONITORING THE EFFECTS OF ADVERSE WEATHER ON CONTRACTS

Since the forms of contract require the contractor to be experienced it would, therefore, be unreasonable for an inexperienced contractor to try to recover costs

## MILLER CONSTRUCTION
### WORKING WEEK SHEET

Duration of Contract ......**24**...... months.    Commencing Date *1st April 1985*
Incidence of Summer/Winter Working ......*2:1*......

| | SITE HOURS | | N.P.O. | | "LOST" TIME | | | | | | EFFECTIVE HOURS (Site Hours less "lost" time) |
|---|---|---|---|---|---|---|---|---|---|---|---|
| | | | | | Summer | | | Winter | | | |
| | Summer | Winter | Summer | Winter | Breaks | Weather | Start & Stop | Breaks | Weather | Start & Stop | |
| MONDAY | 10 hr | 8 hr | 1 hr | – | 20m | 20m | 20m | 20m | 1 hr | 20m | |
| TUESDAY | 10 | 8 | 1 | – | " | " | " | " | " | " | |
| WEDNESDAY | 10 | 8 | 1 | – | " | " | " | " | " | " | |
| THURSDAY | 10 | 8 | 1 | – | " | " | " | " | " | " | |
| FRIDAY | 7 | 7 | – | – | – | – | – | " | " | " | |
| SATURDAY | – | – | – | – | – | – | – | – | – | – | |
| SUNDAY | 8 | 8 | 8 | 8 | 20m | – | 20m | 20m | – | 20m | |
| TOTAL HOURS | 55 | 47 | 12 | 8 | 5.66 | | | 7.33 | | | |
| WEIGHTED AVERAGES | 53 | | 11 | | 6.085 | | | | | | 46.915 |

Notes and Calculations

FIG. 13.1. Example of a Working Week Sheet used to estimate lost time on site due to adverse weather.

incurred through his own inexperience. It is incumbent on the contractor to make reasonable allowances for weather within the contract rates taking the locality and timing into consideration. Once a contract is let, and the contractor makes a start on site, he will normally have a programme approved by the engineer which lays out the order and sequence of the works. The contractor will plan the works on a week by week basis with reference to the overall programme. Most contractors find the most reliable weather forecast for planning the week's activities is the Radio 4 *Farming Forecast* at 13·00 on a Sunday, and repeated on a Monday morning. Using this information he can accelerate or delay weather susceptible activities with that week. The contractor will also keep a retrospective check on the weather comparing the hours lost with the allowances in the tender make up, comparing, for example, long term average rainfall with the actual for that period, and checking five year rainfall averages together with long-term air temperature and wind records.

## EFFECTS OF ADVERSE WEATHER ON CONTRACTS

The contractor insures the works against flooding and storm damage during the construction period and these insurances cover the permanent and temporary works. A vivid example of loss in recent years was a storm in the Loch Lomond area in late July 1986 when over 50 mm of rain fell in a 24-hr period. The contractor was at the foundation stages of building several culverts to take the water off the hillside, under the new road and into the loch. The storm washed down a large amount of detritus which buried the floor panels and the 2 m high starter bars of the wall reinforcement. This incident delayed the contractor several weeks, the cost of which he was unable to recover under the contract conditions. The subsequent repair to the permanent works involved his insurance company in a six figure sum.

A example of exceptionally adverse weather is taken from a road contract recently completed on the west coast. The rainfall for three separate months during 1986 was the amongst the largest on record. Fig. 13.2 shows the day-by-day plot of

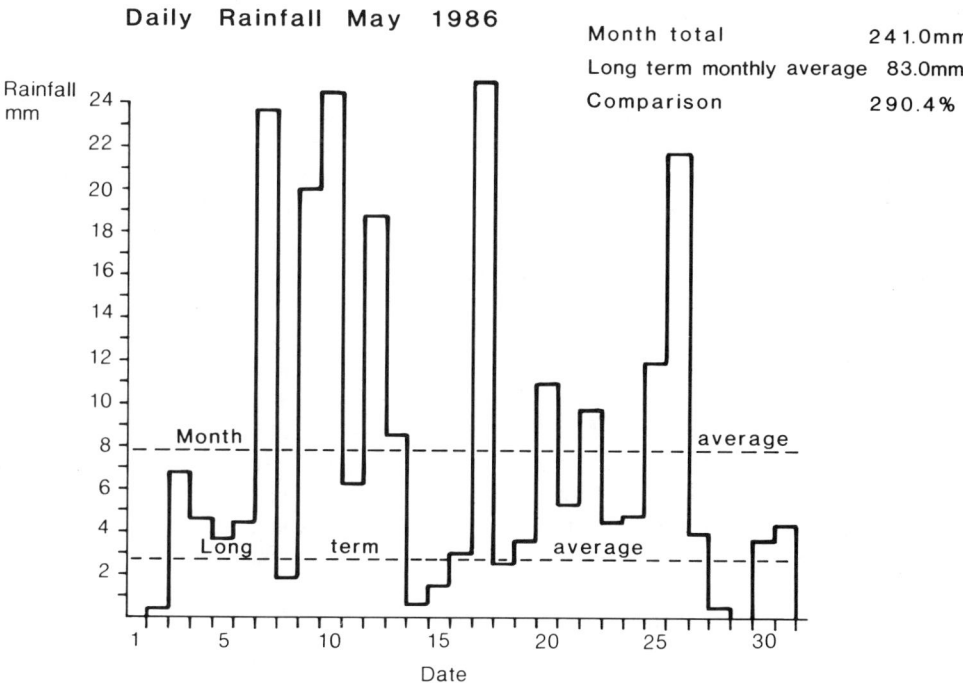

FIG. 13.2. Daily rainfall during May 1986 near a major construction site in Scotland.

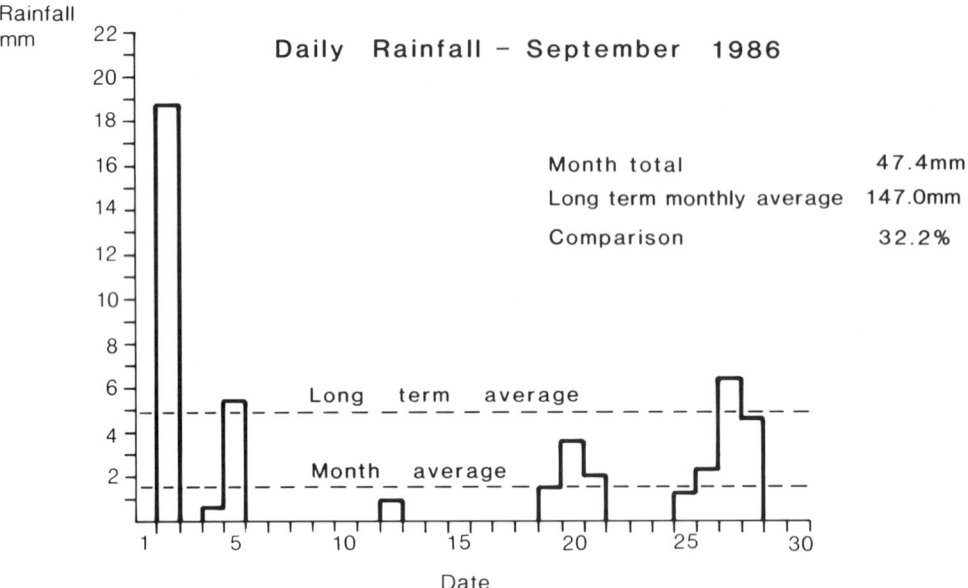

FIG. 13.3. Daily rainfall during September 1986 near a major construction site in Scotland.

rainfall during May 1986 against the long-term average daily fall. From the graph it can be seen that the 7th, 9th, 10th, 12th, 17th and 26th received well above average rainfall. The total for the month was 290% of the long-term average. The contractor would have made little allowance for downtime in his tender as this was supposedly one of the summer months when he could reasonably expect to work unhindered by the weather. The cost of losing these six days of muckshifting was in the region of £60–70,000, for which there was no recovery through the contract. In contrast, Fig. 13.3 shows the daily plot of rainfall during September 1986 against the daily mean. The monthly total was in this case only 32·2% of the long-term average and the contractor lost no time, working within his allowances.

## CONCLUSION

Weather is probably the single biggest cost to both contractor and client in Scotland. Letting contracts with the minimum amount of winter months at the beginning of expected good periods of weather, such as early spring, must be the wisest precaution against incurring additional costs to the contract. The Met Office provides excellent monthly weather records from their various stations. This information, when compared with their statistical data, gives the contractor information that may be useful in his attempt to win back losses due to exceptionally inclement weather. Sites remote from Met Office recording stations should consider installing an automatic weather station, the records from which could be correlated with archived data from long-term stations.

# Forecast services for the building and construction industries in Scotland

## H. CUMMING

*Meteorological Office, Aberdeen*

### INTRODUCTION

WEATHER forecasts for the building and construction industries, like any other forecast, can be considered as a series of mini-forecasts, each of one element, that act together to produce the final forecast. Each element in the forecasts is important for one or more parts of the operational phase of a construction project and the whole forecast has a considerable impact on the progress of the entire project. The elements most significant for the industry, not necessarily in order of importance, are wind speed and direction, temperature, precipitation and humidity.

### FORECAST ELEMENTS AT SITE

*Wind velocity* has an impact on operations that can amount to a work/no work decision. Cranes, particularly tower cranes, have strictly laid down wind limits for operation, which can be enforced by law. The handling of sheet steel and roofing materials can be extremely hazardous in high winds and a prudent foreman must always be on the alert to take men off such jobs before an accident becomes likely. Scaffolders, of course, are in much the same position. These problems are not solely due to the forces exerted by the wind but also to the cooling effect on workers. This cooling can also cause problems with welding.

Wind velocity is directly related to fields of atmospheric pressure which the Met Office's numerical models can forecast very well. An excellent output from the computer, fine tuned by the knowledge of an experienced forecaster, means that wind can be forecast with considerable accuracy for up to 36 hours ahead. Coarser detail, but still fine enough for planning, can be provided up to about three days ahead, and trends can be predicted for about a week.

*Air temperature* has an impact on several construction operations. Mixing, laying and setting of concrete is carried out more efficiently, or perhaps abandoned, with prior knowledge of temperature, while the excavation of earth and its removal are much more difficult with frozen ground. Asphalt also becomes difficult to handle in low temperatures. Welding and jointing may require the creation of a local micro-environment in a tent if the temperature is low. Timber cutting becomes almost impossible if the wood is deep frosted.

Forecasting temperatures, particularly at night, is a delicate operation requiring experience. The temperature can be forecast in moderately fine detail, certainly to within 2°C, for two days ahead with some confidence. Three or four days can be given in coarser detail while trends for a week or more are available.

*Precipitation* as rain, drizzle, snow or hail affects much the same jobs as does air temperature. Earth moving can cost up to ten times as much on a wet day as a dry one. Asphalting needs prediction of rainfall if it is to be undertaken economically and to give a lasting job. Concreting is more effectively done if rain is foreseen and heat loss in welding and jointing work can lead to cutting and rewelding. Rain makes steel slippery and very dangerous to handle.

The timing and amount of precipitation on a site are among the more difficult elements to forecast as they are very much affected by local topography. Showers can be particularly localised. The man-machine mix can, however, provide very useful guidance up to 36 hours ahead, and this would normally be expected to improve as a project progressed and the forecasters became used to site detail. Coarser forecasts are given up to three or four days and generalised outlooks are given for five to six days.

*Humidity* has a limited effect on construction, but where it does have an effect, this can be severe. The major applications are painting and jointing where a high humidity can lower adherence below design tolerance. Humidity is so variable in both space and time that forecasts are difficult to formulate and have to be used with care and understanding. 24 hours is the limit for any attempt at fine detail beyond which, up to four days, it is possible to provide only rather coarse detail and trends to seven days are in very generalised terms.

## MET OFFICE FORECAST SERVICES

The Met Office has established two major centres in Scotland to provide forecast and other services to the public and, in particular, the building and construction industries. In the south, the *Glasgow Weather Centre* serves Borders, Central, Dumfries and Galloway, Lothian and Strathclyde Regions, while *Aberdeen* serves Fife, Grampian, Highland, Tayside and Western Isles Regions. Effectively Scotland is split along a line from the Forth to Oban. Smaller offices in Kirkwall and Sullom Voe, established primarily to look after the oil terminals, provide services to their own communities. There are several ways in which a person with a need for weather advice can obtain it from the Met Office. These include the free public service, premium telephone information services, consultancy services, warning services, dedicated forecasts, and on-site forecasting.

Access to each of the Met Office's forecast offices is available through a telephone number published in phone books. This has the advantage of being cheap, the caller paying only the standard British Telecom charge, and there is no fee accruing to the Met Office. The service is designed primarily to meet the short-term needs of the general public, and detailed advice and guidance will not be provided. For the engineer or project manager this will not be the best alternative. In addition, there will be occasions when the forecaster is dealing with clients' needs and the telephone will be connected to an answering set. Even when the telephone is being answered, the caller with be competing with many others and may need several attempts to get through.

There are a number of recorded telephone weather services available. The best, using forecasts provided by the Met Office, is *Weathercall*. A call to this service is relatively inexpensive and the caller can depend on getting through and receiving information. Forecasts on separate numbers are provided for six areas of the Scottish mainland and islands, conveying in some detail most of the significant variations in the weather though not, of course, providing information specific to individual sites. The messages contain local detail for 48 hours ahead updated three times daily by either Aberdeen or Glasgow Met Offices, with a more general UK-wide outlook for three to five days. While *Weathercall* messages can contain indications of the level of confidence in the current forecast, it is not possible for uncertainties to be discussed with the forecaster.

The Met Office consultancy service allows priority access via an ex-directory telephone number to the forecaster based on a monthly subscription related to the expected usage of the service. For his part, the forecaster learns about the client's situation and problems so that when he answers the telephone he is ready to discuss the weather as it affects the project up to five or so days ahead. The forecaster's prior knowledge of the project and the virtually unlimited discussion and advice make this a very attractive service. Calls are originated by the customer at his convenience.

For those clients who only wish prior warning of the occurrence of adverse weather for their particular operation, the Met Office also offers a warning service. The client specifies the threshold for the weather variables of interest whether this be, for example, temperature, wind speed or snow. The forecaster then undertakes to provide the warning, usually by telephone, as soon as the conditions are forecast. Forecasts are fully site-specific and their provision on monthly subscription allows the forecaster to prepare the warnings with knowledge of the location in mind. This is another relatively cheap service and the responsibility for action rests with the Met Office.

Probably the best forecast service available to the industry is the dedicated forecast, especially where Telex or facsimile facilities are available to the client. At an agreed time, or times, each day the Met Office writes a detailed forecast tailored for the site and project and sends this to the client. Usually the times are chosen so that the client can have a forecast available before work starts in the morning and finishes in the evening. As the forecaster plays a part in setting the times, it can be assumed that he has chosen them to make use of the best data available. He monitors the progress of the weather continuously and sends any up-dates or amendments to the forecast that become necessary. Apart from the telephone discussions that are offered as part of the forecast package, there is a permanent record of the advice upon which the client acted.

The only other method of obtaining forecast advice for a project is to have a dedicated forecaster on the site. The advantages are that the forecaster has a complete understanding of the project and its problems and can supply unlimited advice at any time. Decisions are more easily taken and made with more confidence with a specialist on the team. The forecaster can take the initiative in tailoring his advice rather than simply responding to questions. The whole weather aspect of the project is under constant review and engineers can be helped and prompted in their decision making. This service is expensive and can probably only be justified for large projects, where weather-sensitive decisions have to be made requiring detailed on-site knowledge of the conditions.

# Weather interference with construction operations: Met Office climatological services

## M. J. PRIOR

*Meteorological Office, Bracknell*

## INTRODUCTION

WHEN devising meteorological services for the construction industry it is essential to be aware of the ways in which the weather can affect operations in terms of working practices, contractual obligations, and health and safety requirements. Climatological services are used at the tendering, planning and project monitoring stages of a contract, and weather forecasts are required during the operational stage. In this way the best use is made of the weather information available. This paper outlines the applications of past weather data to the construction industry.

Losses of time due to bad weather have long been of concern to the building and civil engineering industries, and fall into two categories:

(a) Delays when adverse weather stops work or reduces working efficiency. These may result from handling limitations, including those concerning safety, comfort, and efficiency, and the avoidance of damage to equipment. They may also be due to materials limitations associated with the need to achieve a certain quality and durability.

(b) Delays caused by damaging weather events such as gales, prolonged rain and deep snow. Since these are relatively rare, they do not normally need to be allowed for when tendering. However, precautions still have to be taken to limit the vulnerability of the work at critical times during the contract.

## COSTS TO THE CONSTRUCTION INDUSTRY

The average proportion of hours lost on construction sites because of bad weather has been estimated from a study of 39 sites as ranging from 7% of all site hours in Scotland to 3% in south-east England (Smith and Rawlings, 1974). Thus the national cost of such stoppages runs into many hundreds of millions of pounds each year. An update of earlier estimates is about £350–400 million in a typical year. It should be noted that these losses relate to both external and internal operations. Those trades that are most vulnerable to bad weather, such as roofing contractors, will lose proportions of possible working hours several times greater than the proportions of all site hours. The worst delays occur in winter in Scotland and upland areas elsewhere in the United Kingdom. For example, it has been estimated that, on average, a quarter of working time is lost between December and February in Glasgow because of precipitation intensities greater than $0.5 \text{ mm hr}^{-1}$ or temperatures below 1°C (Brown, 1961).

M. J. Prior

# CLIMATOLOGICAL INFORMATION AS AN AID TO DECISION-MAKING

There is a considerable archive of climatological data held by the Met Office (Singleton this volume). These data may be used at the stages of tendering, long-term planning of site operations after a contract has been awarded, and retrospective analysis of weather experienced during a contract. At the tendering and planning stages the averages and likely variability of relevant weather elements will be required, representing the climatic conditions of the contract site. Later, a climatological interpretation may be needed of the weather experienced, especially if it was adverse. Such interpretations are frequently needed as evidence in relation to claims for extensions of time. The challenge to the climatologist is to make the information provided as representative as possible of the construction site and the work being undertaken there.

The UK network of official weather stations comprises several types, including those recording hourly throughout the day. For the construction industry, whilst daily information concerning, for example, frost, rainfall amount and snow depth, is useful for some applications, the principal need is for hourly information so that the weather during working hours may be assessed. The spatial density of weather stations, particularly hourly ones, is greatest in the more populous parts of the UK. In Scotland, this means that, whereas the Central Lowlands and coastal areas are reasonably well served, there is only a sparse network in many of the upland areas. The requirement for site-specific climatological data may sometimes be met simply by choosing the most representative weather station. When it is clear that none is available, a less direct method needs to be adopted. Non-availability is likely to be more common in Scotland because of its complex topography.

The use of climatological maps is well established. These range from long-term averages that may be used to help interpret data recorded at weather stations, to those especially prepared for tendering and planning purposes. One example of the latter type is the set of maps used for the CLIMEST service, introduced in 1969. This provides seven sets of monthly averages relating to rainfall, temperature, and windspeed; for example days with air frost and hours with wind gusts exceeding $17\,\text{ms}^{-1}$. Following recommendations by Smith and Rawlings (1974), improved computing capabilities have been used to prepare maps with more emphasis on the weather during working hours, which are taken to be 0700 to 1700 GMT.

The techniques used to prepare *working day* maps usually involve correlating the information that the contractor needs, which may be calculated at the hourly observing stations (for example, average number of hours per month with air temperature below 2°C), with information that may be mapped because it is available for a denser network of daily and hourly stations. For example, King (1981) used detailed maps of average annual rainfall amount to produce a set of maps of working day rainfall in terms of rainfall duration, wet days and dry days. Although the CLIMEST service itself has not yet been extended or modified, when a contractor's need is best met by estimated averages then the full range of maps available is used.

New technology increasingly offers the chance of using other methods of making information more site-specific. Among the ones currently available are:

(a) The expansion of the weather station network through the deployment of more automatic recorders, providing continuous information about those elements of most concern to the contractor which include air temperature, wind speed, and rainfall. On-site records are also possible using data loggers.

(b) The use of radar observations of rainfall to interpolate between the measurements made at rainfall stations. A system has recently been established for the routine processing of daily data from the network of six radars currently in use (May, 1988).

(c) The use of theoretical and empirical models to predict the influence of topography and ground roughness on the climate of the contract site, usually in comparison with the known conditions at a weather station in the same area. Current examples include the assessment of wind conditions using the STRONGBLOW microcomputer package (Cook et al., 1985) and of the incidence of low temperatures in valleys (Tabony, 1985).

Comprehensive descriptions of the effects of weather on construction processes are given by Lacy (1977) for the UK and by Russo (1971) for the USA. Recent work by the Met Office and the Building Research Establishment has concentrated on the identification of appropriate stop-work thresholds for a wide range of building and civil engineering operations. This has been done mainly by a literature search, including British Standards Institution *Codes of Practice,* Department of Transport specifications, and research association and trade federation publications. The initial findings were used in a paper concerning weather forecasting for construction sites (Prior and King, 1981) and were confirmed indirectly by Smith (1983) in terms of observed trends in demand for weather forecasts.

A more comprehensive study of weather thresholds has recently been carried out in the form of a pilot guide describing the choice of thresholds for eighteen operations accompanied by monthly analyses of the time these are exceeded at Plymouth (Keeble and Prior, 1988). The operations considered in this guide are listed in Table 15.1 together with the meteorological elements thought to affect them. Some of the operations are sub-divided to reflect the common alternative working methods such as the surface dressing of roads, which is discussed in terms of both hot tar and bitumen emulsion bonding. The guide discusses the meteorological information required in terms of specific stop-work thresholds. For example, concreting horizontal members is deemed to be affected when either the air temperature is below 2°C or there is rain falling at a rate of at least $0.5$ mm hr$^{-1}$. The chosen thresholds are meant to represent the point at which a typical contractor will stop work. Clearly, if he is better equipped and organised to deal with adverse weather, the thresholds may be too conservative. As well as the specific operations listed in Table 15.1, the influences of availability of daylight, comfort and dexterity, mobility and damage are dealt with separately.

TABLE 15.1. *Meteorological information requirements for construction operations*

| Operation | Precipitation | State-of-ground* | Wind speed | Air temperature | Visibility | Humidity |
|---|---|---|---|---|---|---|
| Surveying and setting out | ✓ | | ✓ | ✓ | ✓ | |
| Excavation and earth moving | ✓ | ✓ | | | | |
| Use of cranes | | | ✓ | | ✓ | |
| Lifting wind-sensitive loads | | | ✓ | | ✓ | |
| Erecting frameworks | ✓ | ✓ | ✓ | | ✓ | |
| Suspended access equipment | ✓ | . | ✓ | | | |
| Welding | ✓ | | | (wind chill) | | |
| Concreting | ✓ | | | ✓ | | |
| Removing formwork/temporary props | | | | ✓ | | |
| Brickwork/blockwork | ✓ | | ✓ | ✓ | | |
| Rendering | ✓ | | ✓ | ✓ | | |
| Rolled asphalt paving | ✓ | ✓ | | (wind chill) | | |
| Surface dressing | ✓ | ✓ | | ✓ | | |
| Asphalt: roofs/bridges/tanking | ✓ | ✓ | ✓ | | | |
| Membrane roofing | ✓ | | ✓ | ✓ | | |
| Tiling, slating, sheeting | ✓ | ✓ | ✓ | | | |
| Glazing | ✓ | ✓ | ✓ | ✓ | | |
| Painting, joint sealing | ✓ | | ✓ | ✓ | | ✓ |

\* 'State of ground' codes record whether the ground is moist, wet, icy, snow-covered etc and may also be used to indicate whether materials or components are in a suitable state for use.

The analyses given in the guide are based upon hourly data recorded at Plymouth from 1957 to 1981. They relate to the numbers of clock hours affected during a fixed working day of 0700–1700 GMT, all days of each month being considered. Thus, they are intended to reflect adverse weather at the time of the operations, this having the most effect. Some operations are also subject to delayed effects, such as residual surface wetness after overnight rain, or anticipated effects such as subsequent damage by frost. The use of the *state of ground* code to some extent reflects the former. There were various choices available for analysing the recorded hourly observations (air temperature, humidity, visibility), hourly means (wind speed), hourly totals (rainfall duration and amount) and 3-hourly observations (state of ground). Precipitation perhaps presents the most problems, being discontinuous in nature. Stoppages on some operations clearly coincide with the duration of precipitation whereas on others they exceed it, because of wet materials or surfaces.

## Weather interference with construction operations

For many of the operations in the guide, if the mean intensity during the hour exceeded the stop-work threshold then the whole clock hour was deemed to be unsuitable.

## EXCEPTIONALLY ADVERSE WEATHER

When tendering, the contractor needs to assess the risk of adverse weather. The location and the seasons during which different phases of the work are due to take place have an obvious influence. If weather risks are to be appraised properly, account must be taken of the year to year variability of weather delays and not just

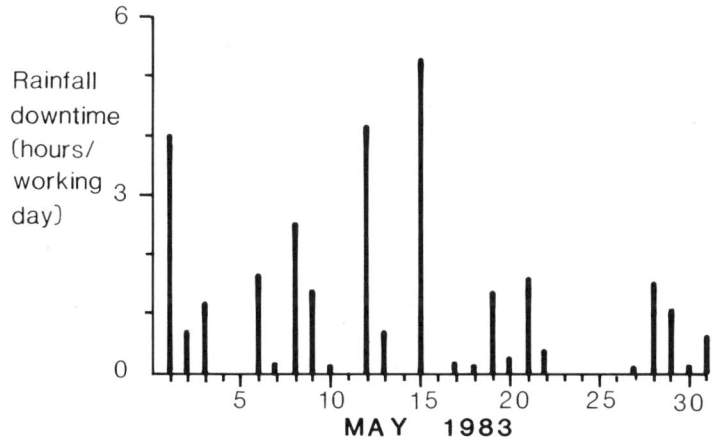

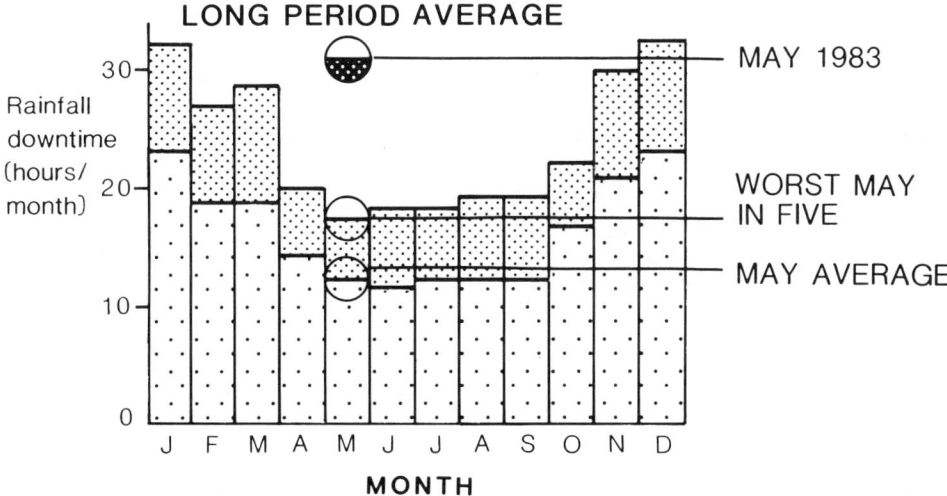

FIG. 15.1. Assessment of working hours lost due to rainfall at a construction site during May 1983 in relation to the long-term average.

of average values. The basis of the tender may then be used for comparison with the weather conditions experienced on site when monitoring progress (Carson this volume). The most widely used forms of contract allow an extension of time if the contractor experiences 'exceptionally adverse weather'. This certainly suggests something substantially worse than the average and for several years the Met Office has been advocating a comparison of contract weather not with average conditions but with those typically occurring once in 5 years. This criterion has been used for individual enquiries but, despite some publicity, its acceptance generally has been slow. However, it has been adopted by the Scottish Special Housing Association and, provisionally, by the Royal Incorporation of Architects in Scotland for deciding adverse weather extensions under the Joint Contracts Tribunal form.

When an analysis of adverse weather is carried out, the value of affected working hours exceeded typically once in 5 years is the upper quintile and the value not reached once in 5 years is the lower quintile. Thus the average, upper and lower quintiles represent normal conditions and the limits of *exceptionally adverse* and *very favourable* conditions respectively. Providing a long enough record is available (at least 25 years preferably), it is possible to calculate these for daily data (for example, number of frosts or days with snow lying), or hourly data (for example, working hours with adverse weather) on a monthly basis. This format has been adopted in the Plymouth guide (Keeble and Prior, 1988). An example of their use when assessing a particular month is given in Fig. 15.1 which is taken from '*Services to the Construction Industry*' (Met Office, 1985). This example relates to a contract for which the total number of wet working hours lost in May 1983 was over 30. This was not only more than the long term average but also greatly exceeded the number of hours lost in the worst year in five.

## CONCLUDING REMARKS

A description has been given of the uses of climatological data analysis techniques at the tendering and planning stages of a contract and when assessing progress on site, particularly if weather conditions have been adverse. The benefits for the contractor in so doing lie not only in improved planning but in the swifter resolution of any claims which may arise.

ACKNOWLEDGEMENT. The author would like to record his appreciation of Eric Keeble of the Building Research Establishment for his contribution to the research upon which much of this paper is based.

# Weather sensitivity in the gas industry

R. A. STEEL

*(British Gas plc, Scotland*

## INTRODUCTION

IN recent decades, gas has tended to replace more traditional fuels for maintaining the comfort levels of the Scottish population. Domestic appliances are now technically advanced and react quickly to the changes in environmental conditions. Most central heating systems start up when the customer is absent from the premises or asleep. Fuel consumption is determined by temperature measurements of the living areas which are influenced by the outside weather. In the industrial and commercial field, the process use of gas is controlled by sophisticated apparatus whilst the space heating load has become very sensitive to variations in ambient conditions through the advent of new energy management systems.

The gas industry must, therefore, maintain a continuous supply of fuel which has a highly weather sensitive demand. To do this securely and economically requires an awareness of weather parameters, together with prime costs, speed of delivery and storage facilities. In the long term, there is a requirement to negotiate supplies from the North Sea years ahead. To help determine future demand, weather records are used in the context of each winter's experience to define the network capacity under 1:20-year severe weather conditions. On the short time-scale, gas must be ordered at 16.00 hours for the following day (06.00–06.00). If the rate of supply needs to be changed, this can only be increased by up to 5% at two hours notice and demand can fluctuate widely. For example, the industry can be required to provide six times the daily summer demand on a peak winter day.

Weather data are used in two main ways:
1. through determination of the relationship which links gas demand to weather.
2. by application of previous weather experience, combined with this relationship, to provide models which guide management decisions to a cost-effective and secure conclusion.

Over the years, the increasing sensitivity of space-heating requirements to weather conditions has produced a much more detailed understanding of the relationships between gas demand and meteorological factors. While it may seem logical and obvious that the daily demand for gas depends on temperature, it is not quite so obvious what measure of temperature should be used.

## EFFECTIVE TEMPERATURE AND GAS DEMAND RELATIONSHIPS

The ambient temperature varies temporally and spatially across Scotland. What is needed is a single value which is representative of the demand for gas over the whole country for an entire day. After research into many locations and composite values, temperature data from Springburn Park, Glasgow was chosen as the best thermal index of gas demand in Scotland. It was also found that the simple mean temperatures ($\frac{1}{2}$ max $+ \frac{1}{2}$ min) gave a better fit than weighted values such as $\frac{2}{3}$ max $+ \frac{1}{3}$ min. In addition, it is known that gas demand lags behind temperature changes. This means that today's demand depends not only on today's temperature but is also influenced by conditions on previous days. This effect is incorporated into the daily "effective temperature" (Teff), defined in terms of the mean temperature (Tm) on any day as

$$\text{Teff}_i = \tfrac{1}{2} \text{Tm}_i + \tfrac{1}{4} \text{Tm}_{i-1} + \tfrac{1}{8} \text{Tm}_{i-2} + \tfrac{1}{16} \text{Tm}_{i-3} + \cdots.$$

It can be shown that this is equivalent to

$$\text{Teff}_i = \tfrac{1}{2}(\text{Tm}_i + \text{Teff}_{i-1}).$$

Using effective temperature, a straight line fitted to the data for a sample year would give a reasonable fit. This is illustrated in Fig. 16.1(a), where the slope of the line is 147 thousand therms per degree Centigrade. However, as in other years, the straight line over-estimates demand at the colder end of the range. If the data are confined to a single month, the relationship will differ significantly. This can be seen in Fig. 16.1(b) for October and February where the slopes of the lines are 95·2 and 111·8

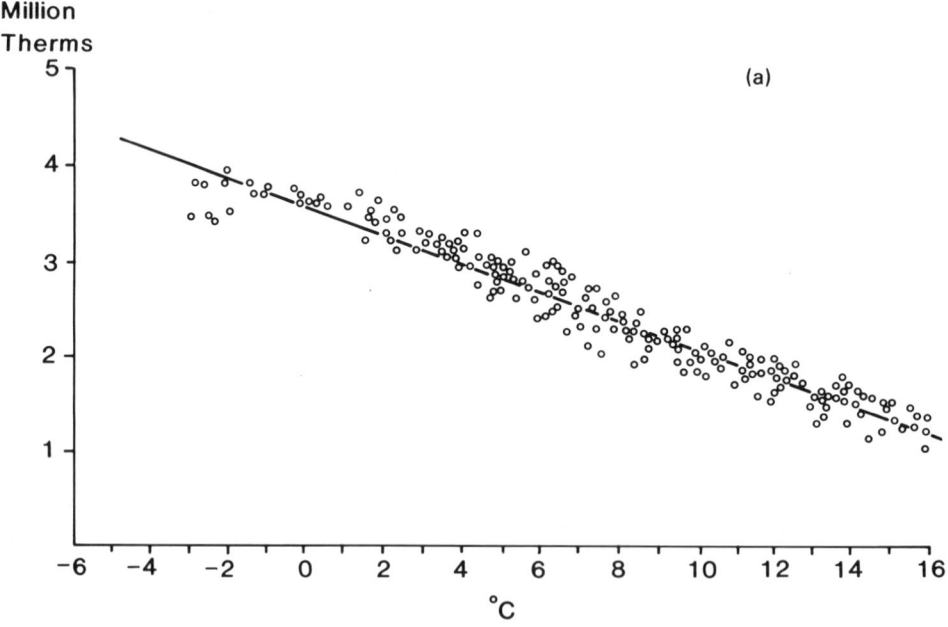

FIG. 16.1. Daily gas demand related to effective temperatures 1977–78. (a) represents the overall annual relationship; (b) shows variations from the annual fit for two sample months; (c) shows the improved fit at lower temperatures.

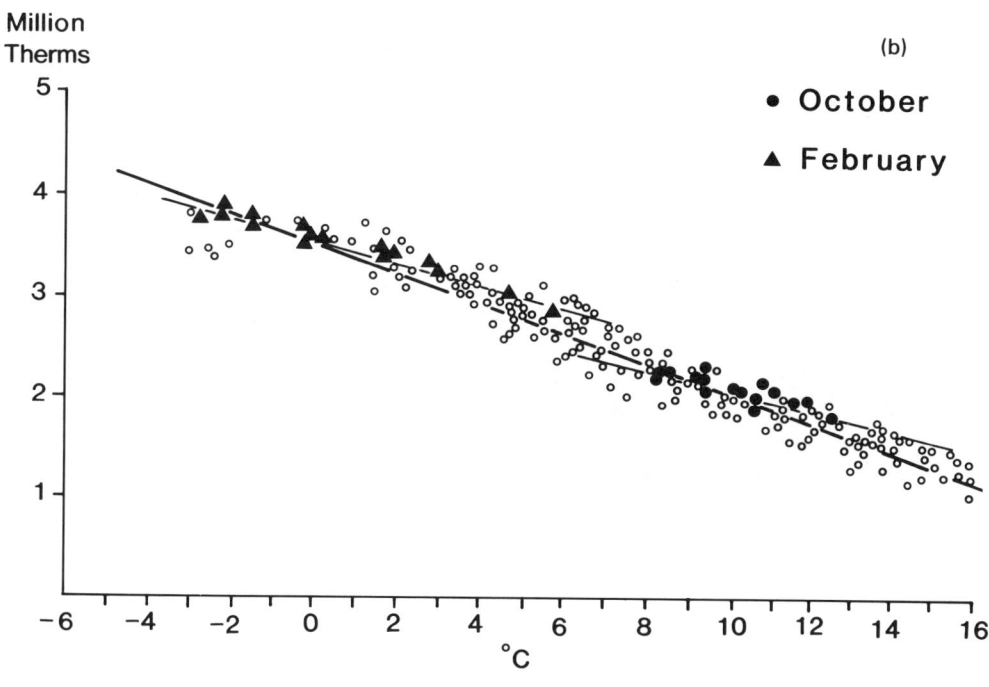
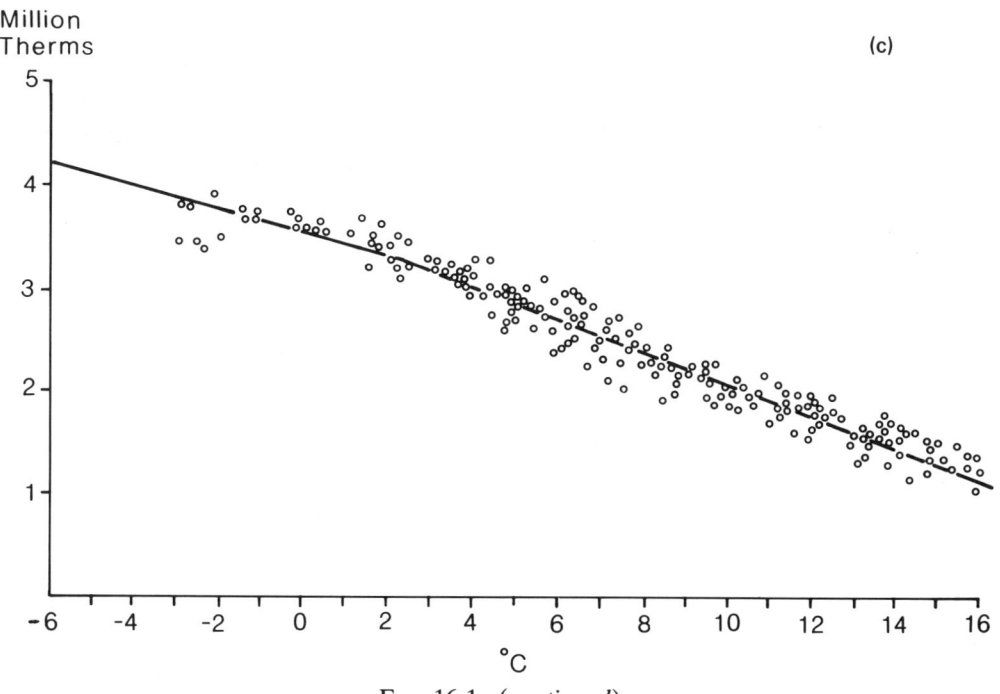

Fig. 16.1. (*continued*)

respectively. Indeed, the slope of the line drawn for any month in 1977–78 is less than the slope of the line fitted to the year as a whole and the individual monthly regression lines appear closer to each other than to the annual line.

Such results suggest the existence of short-term behaviour which differs from the underlying relationship. The long-term description of temperature is incorporated within a collection of 366 numbers, together known as the "seasonal normal temperature pattern" (TSN). Each of these is essentially the simple arithmetic mean of the effective temperature experienced on the same day over the last 59 years, which is the standard weather database period for all British Gas regions. This value is held to be the effective temperature most likely to occur on that day in future years.

It can be demonstrated that the demand for gas is a combination of a seasonal effect together with an effect which represents the shorter term response to weather fluctuations around that seasonal norm. Algebraically the season is represented by the seasonal normal temperature and fluctuations as the difference between this and effective temperature:

$$Y_i = a + b\text{TSN}_i + C(\text{TSN}_i - \text{Teff}_i) + E_i$$

where $Y_i$ = demand for gas on day $i$
$\text{TSN}_i$ = seasonal normal temperature on day $i$
$\text{Teff}_i$ = effective temperature on day $i$
$E_i$ = random error term

Graphically the third term of above equation $C(\text{TSN}_i - \text{Teff}_i)$ does not show distinctly until the temperature drops below the seasonal normal temperature range (2·5°C). At that point, there is a flattening of demand as the seasonal variation is inoperative. For this reason, together with the great interest in demand during the coldest weather, the above relationship is usually depicted by the kinked line in Fig. 16.1(c).

## REVISION OF THE BASIC DEMAND MODEL

The gas demand model described above was first adopted by Scottish Region in 1977 and soon became the basis of the method recommended by British Gas to all regions. Since then, ongoing research has revealed the need for two major overhauls of the basic model. The first of the changes involved the addition of a wind chill element. This was required as a result of experience in February 1979 when strong winds combined with low temperatures and driving snow for three consecutive days to create demand much higher than expected. It is believed that such increased demand arises because the cold airflow strips away layers of warm air surrounding individual houses and also promotes physiological cooling of the occupants.

Although it was clearly necessary to introduce a wind-related factor into the

## Weather sensitivity in the gas industry

equation, evidence from previous years showed that this factor must only operate when both temperature and wind conditions are extreme. To approximate this, a wind threshold was defined below which wind would be assumed to have no effect and a similar temperature threshold was defined above which temperature was deemed to have no effect. These thresholds were determined by trial and error to be 10 knots and 3°C respectively.

The model which gave the best fit for wind chill was:

$$y_i = a + b\,\text{TSN}_i + C(\text{TSN}_i - \text{Teff}_i) + d(\text{W} \cdot \text{DD})_i + E_i$$

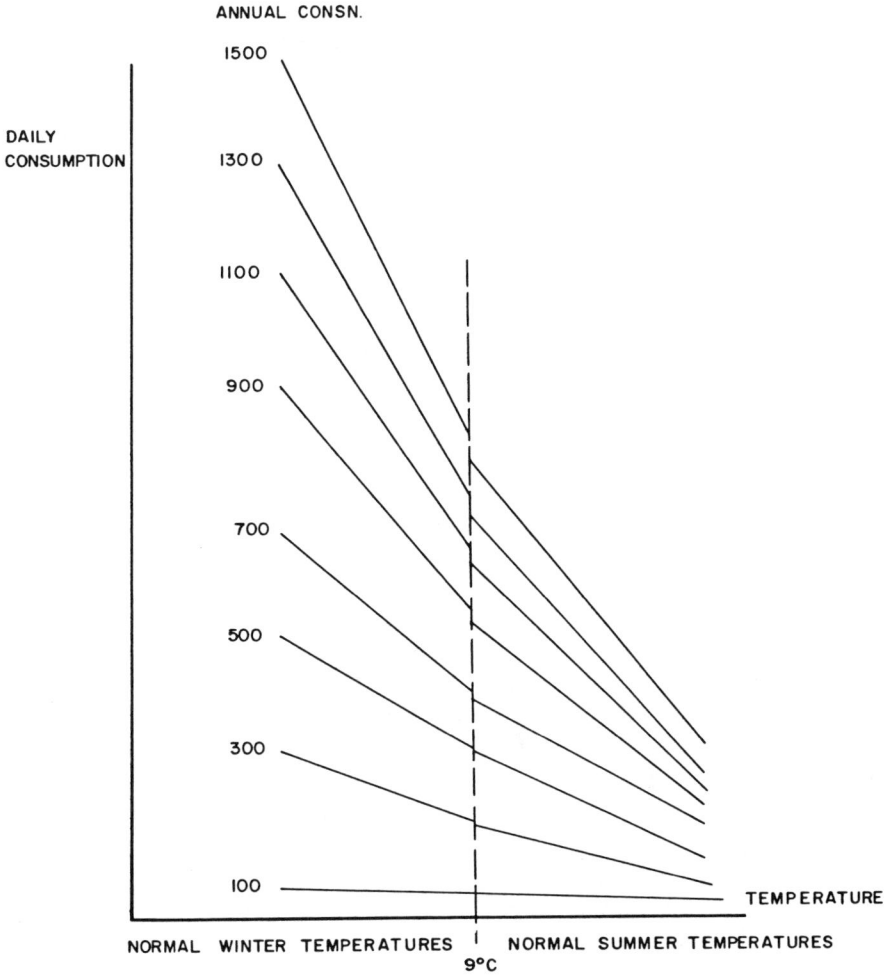

FIG. 16.2. Domestic temperature-demand relationships for different annual consumption bands showing switch on/switch off effects.

where

$Y_i$ = demand for gas on day $i$
$a + b\,\text{TSN}_i$ = demand according to time of year on day $i$
$C(\text{TSN}_i - \text{Teff}_i)$ = demand according to relative temperature for the time of year
$d(W_i \cdot \text{DD}_i)$ = demand caused by combined wind and temperature, where
$W_i$ = number of knots above 5 knots
$\text{DD}_i$ = number of degrees less than 5°C
$E_i$ = random error term

Although this relationship appears to work reasonably well, it is difficult to estimate because of the small number of occasions when sufficiently severe weather conditions combine to produce data for analysis.

The second modification to the basic demand model has been necessary to reflect the fact that domestic customer behaviour is different during the summer period from the winter. This is known as the summer switch-off. It is perhaps best illustrated by looking at domestic demand-temperature relationships analysed by consumption bands, as in Fig. 16.2. For a low consumption customer the summer relationship is not statistically different from that for winter. In the upper consumption bands a small discontinuity appears at about 9°C and by 700 therms per annum (the size of a smallish central heating system) the discrepancy is quite distinct. An appreciation of the switch-on/switch-off division of the supply year into two parts allows a more realistic analysis of past data and, therefore, a more accurate set of projections.

## LONG-TERM PEAK DEMAND FORECASTING

Given forecasts of annual sales for the domestic, industrial and commercial market sectors, it is necessary to ensure that the gas supply and distribution network is capable of performing under the most severe weather conditions. In practice, this means forecasting demand for a day so severe that it would occur only in one year out of 20 and for a period of days that would only occur once in a 50-year period.

The method of calculating peak days involves taking the highest demand from each of the 59 years of weather record. In Fig. 16.3 these peak values are plotted in ascending order with a Gumbel–Jenkinson curve fitted through them. From this curve, the 95th percentile, or 1 in 20 level, can be read off. These calculations are performed with a computer model which also permits simulation of the unexplained scatter of points around the best fit line. The model deals with this by simulating each of the 59 years 28 times, each time with random errors which model the real-life variability of daily demand, even when the temperature and wind remain the same. The simulation model calculates the 1 in 20 demand from each of its 28 simulations and takes the average of these as being the best estimate of the true 1 in 20 peak day.

The derivation of extreme demand years relies on the use of load duration curves. A load duration curve is simply a graph of a set of ranked daily demands over a

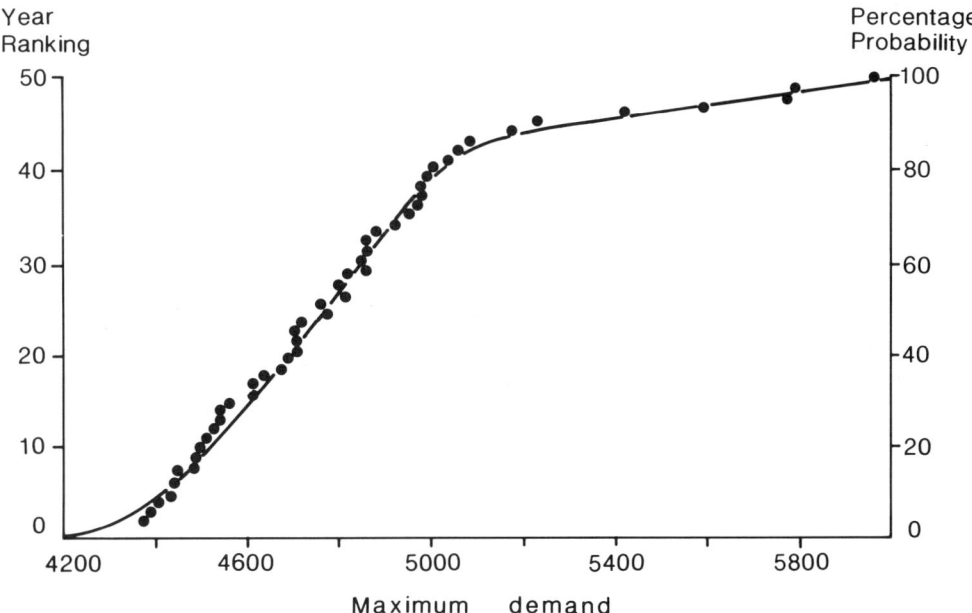

FIG. 16.3. A Jenkinson curve fitted to the ranked maximum gas demands in order to quantify a 1 in 20 peak day.

supply year, from highest to lowest, regardless of the order in which they occur. The most commonly used load duration curve is the 1 in 50 curve, which has to be determined for each winter of the planning period. As an example, the actual weather for the period 1928–29 to 1986–87 is substituted into the demand-weather equation. This gives a set of 59 years of daily demands which would occur if the weather of 1928–29, 1929–30 etc was repeated in 1986–87. From each year's demands a load curve can be constructed.

One method of analysing and comparing these load curves is to express them in terms of volumes of gas above specified thresholds. If the thresholds are taken one at a time, then the 59 values of volume of demand above that threshold for the years 1928–29 to 1986–87 can be examined. It has been found that the cube roots of these volumes follow a normal distribution. The 1 in 50 level is the point at which 98% of the values lie below the point and 2% lie above it. By fitting normal distributions to the cube roots of volumes above a wide selection of thresholds for each of the 59 years, 1 in 50 values can be found for each threshold. It is then a fairly straightforward matter to reverse the process and derive the 1 in 50 load curve from the volumes above each threshold by differencing.

## FORECASTING STORAGE REQUIREMENTS

Internal storage in the form of gasholders and linepack is necessary to allow the demand for gas within a day to be matched by the supply. According to national operating rules, gas regions should aim to take supplies from the National Transmission System at a constant rate per hour, using regional storage to supplement the supply at periods of high demand within the day. The internal storage is then filled up again when demand is low. Fig. 16.4a shows a typical within-day demand profile and the corresponding flat supply rate. The shaded area represents the volume of regional storage required to balance supply and demand. This volume is usually expressed as a percentage of the day's demand and is called the diurnal swing. The peak day diurnal swing in the base year is estimated from the relationships between observed diurnal swings and actual demands. This is projected to peak levels to give an estimate of what the peak day diurnal swing would have been. From a knowledge of the expected growth of different market sectors and their diurnal patterns, estimates of future peak day diurnal swings are made. In reality, forecasting the daily demand is subject to error, so the supply is not taken in straight line but in steps, as shown in Fig. 16.4b. In this case the demand is under-forecast at the start of the day. As the day progresses, the forecasts improve

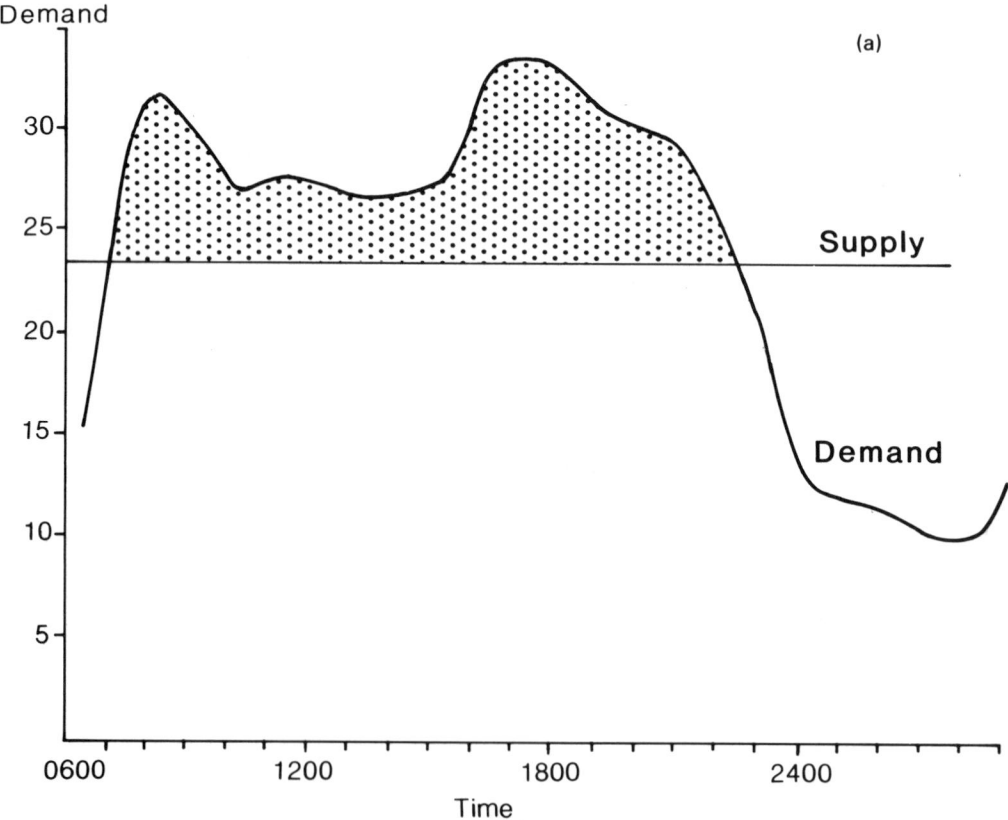

FIG. 16.4. (a) A typical within-day gas demand profile with theoretical constant supply; (b) The same demand profile showing a normal stepped supply pattern.

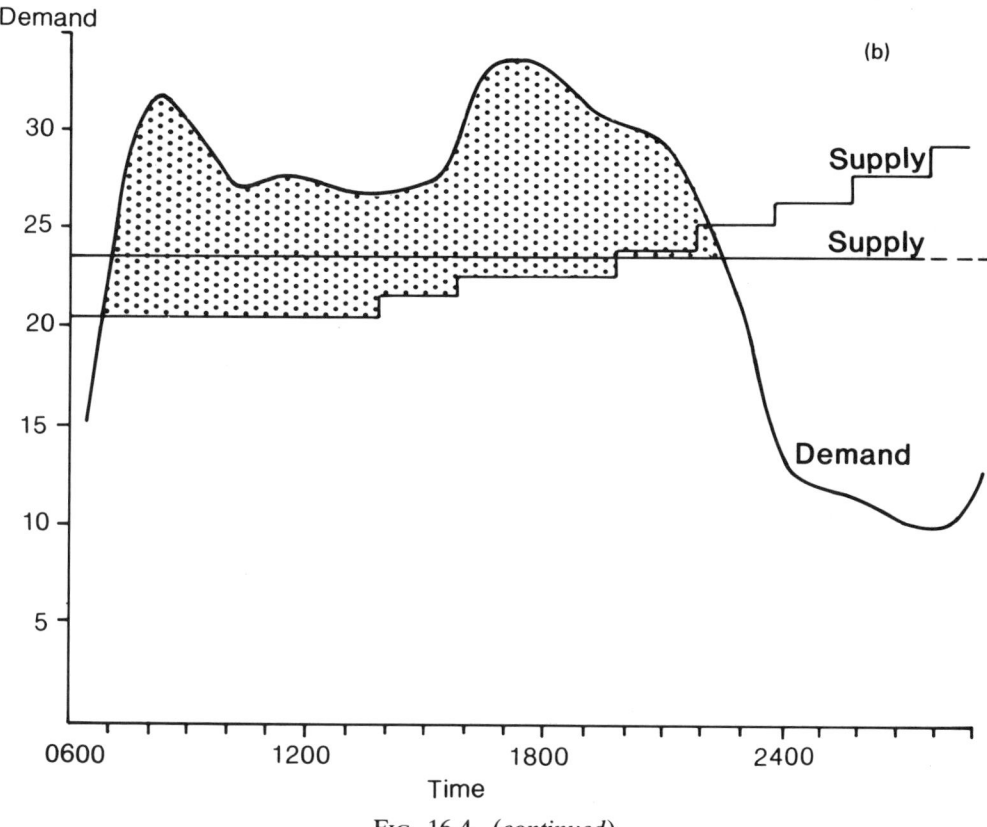

Fig. 16.4. (*continued*)

until supply eventually matches demand. However, extra internal storage was required because of the forecasting error and this is represented by the increased area of shading on Fig. 16.4b.

A computerised simulation model has been written to help in forecasting the internal storage requirements. In this model the operations of the Grid Control room are simulated. The model is run for a given year, say 1988–89. The weather for each year within the 59-year database is used to give forecasts of 1988–89 demands under the weather conditions of each year. As the program works through each day of the 59 years, it calculates an hourly demand profile using a selected diurnal swing which is typical for the time of year. Every four hours, the model makes a forecast of the demand for the day in progress. These forecasts are subject to errors which are based on the actual levels of forecasting error which occur in practice. On the basis of these forecasts gas intake rate is altered, interruptible customers are interrupted and other operations exist, exactly as in real life. From the resulting demand and supply profiles, a storage profile is calculated and the minimum internal storage required is determined.

R. A. Steel

# GAS PRESSURE AND INTERRUPTIBLE CUSTOMERS

A consequence of the seasonality of gas demand is that pressure in the grid network must be increased to cope with higher demands in winter. Until 1978 the pressure switching occurred on 1 October each year. This achieved security of supplies but entailed operating in most years for an unnecessarily long period at the higher pressures with greater leakage losses and compression costs. Now policy is based on a switch temperature which has permitted some considerable delays in switching to the winter setting. Indeed, since 1978 switch dates have never been earlier than 28 November and have sometimes been as late as mid-January.

Weather information is also used in connection with interruptible customers. These are customers who, for a financial consideration, agree to a contract which allows supplies to be withdrawn on a specified number of days per year following a telephoned request from British Gas. In order to operate these contracts, a scheduling system was introduced. This entails processing weather information each week such that enough interruption is reserved to meet the worst winter that has been experienced from that date onwards and declaring the rest available for use. The interruption thus scheduled for the coming week is given an interruption probability coupled with a critical temperature. This ensures that the customer is prepared for interruption for a minimum period.

TABLE 16.1. *Information requirements for gas operations*

Telex received at 08.00, 12.00, 15.30 and 24.00 hours each day

| Information received | 08.00 hrs | 12.00 hrs | 15.30 hrs | 24.00 hrs |
|---|---|---|---|---|
| A | X | X | X | X |
| B | X | X | X | X |
| C | X | X | X | X |
| D | X | X | X | X |
| E | X | X | X | X |
| F | X | X | X | X |
| G | X | X | X | X |
| H | X |   | X |   |
| I | X |   | X |   |
| J | X |   | X |   |
| K |   |   | X |   |
| L | X |   | X |   |
| M | X |   | X |   |
| N | X |   | X |   |
| O | X |   | X |   |

(*See page* 159)

## WITHIN-DAY FORECASTING

The Grid Control office at British Gas Scotland headquarters receives information from the Met Office concerning both actual and forecast weather conditions to help short-term scheduling of gas supplies. This information consists of data from Abbotsinch and Aberdeen, as shown in Table 16.1. Much of the information relates to two-hourly temperatures, windspeeds and rainfall for Abbotsinch received at regular four-hour intervals. In addition, Grid Control also know the cumulative gas send-out for the current day and this value is updated hourly. From this information, forecasts of the gas demand for both today and tomorrow have to be made and continuously updated as conditions change. Regression analysis of hourly

---

Key

A   Reference Point
B   Date
C   Time of Issue
D   Actual Temps (deg C)—3 hours before time of issue and 1 hour before time of issue. Also at 08.00 hrs and 24.00 hours additional actual temps (deg C) for 7 hours and 5 hours before time of issue.
E   Forecast temps—(deg C) 1 hour after time of issue and every 2 hours for next 24 hours.
F   Actual wind speed and direction at time sent—knots
G   Forecast wind speed and direction—knots
H   Maximum temperature for day 2 (deg C)
I   Minimum temperature for night 2 (deg C)
J   Maximum temperature for day 3 (deg C)
K   Minimum temperature for night 3 (deg C)
    Maximum temperature for day 4 (deg C)
L   Forecast in general terms of cloud, sun and precipitation for first 24 hours.
M   Comments in general terms for temp. and whole forecast period.
N   Forecast in general terms for wind for second 24 hours.
O   Forecast in general terms for cloud, sun and precipitation for second 24 hours.

The following information is obtained each morning by telephone from Aberdeen and is available from 06.30 hours (covers 20 mile radius of Edinburgh):- Period covered–next 24 hrs. Barometer Windspeed-mph Forecast Direction Further outlook Temperature (deg C) Temp. (deg C) at 14.00 hrs next day.

gas send-outs gives the within-day estimate of demand each hour from 07.00 to 05.00 hours, with the facility to produce either end-of-day values only or a full hour by hour interpretation. The estimated outputs have been found to compare favourably with the actual supply over the last twelve years.

One method gives the estimate of gas send-out for the current day at four hourly intervals, commencing at 08.00 hours to coincide with the receipt of temperature forecasts from Abbotsinch. A regression technique is employed whereby the coefficients are calculated at the start of every Grid Day by taking the end-of-day sent-outs, and effective forecast temperatures for today of the $N$ most recent historical days and fitting them to an equation of the following type:

$$Y_0 = Y_{-1} + b + a\,\Delta T_H$$

where

$Y_0$ = the forecast send-out for today
$Y_{-1}$ = the actual send-out for the last similar day type
$\Delta T_H$ = the change in effective forecast temperature between today and the last similar day type made at time H
$a, b$ = regression coefficients
$N = 20$ days for weekdays (Mon–Tues–Wed–Thurs)
$N = 10$ days for weekend days (Fri–Sat–Sun)

The effective temperature used is derived from hourly temperatures in the following way:

$$T_E = 0{\cdot}6 T_\Delta + 0{\cdot}4 T_{E-1}$$

where $T_E$ = effective forecast temperature for today
$T_{E-1}$ = effective forecast temperature for last similar day type

and

$$T_\Delta = \frac{0{\cdot}75(T_1 + T_2 + \cdots + T_8)}{8} + \frac{0{\cdot}25(T_9 + T_{10} + T_{11} + T_{12})}{4}$$

where $T_1 \rightarrow T_{12}$ are today's two hourly actual or forecast temperatures. The weighted temperature $T_\Delta$ is biased towards the first eight temperatures as they occur during the so-called 'living day'. The last four temperatures $T_9 \rightarrow T_{12}$ cover the period 23.00 to 05.00 hours when the load is less temperature sensitive.

A second method gives the day-ahead estimate of gas send-out, again at four-hourly intervals commencing at 08.00 hours. This model is very similar since it also uses a regression technique relating effective forecast temperatures to send-outs. The forecast temperatures received for tomorrow differ from those received for today at 08.00 and 12.00 hours. For these hours there are fewer forecast temperatures provided and the decision was made to use the maximum and minimum temperatures as the basis for calculating the effective forecast temperatures at these particular times of day.

The effective temperature becomes:

$$T_E = 0.6 T_\Delta + 0.4 T_{E-1}$$

where $T_E$ = effective temperature for tomorrow
$T_\Delta$ = weighted forecast temperature for tomorrow such that
$T_\Delta = 0.6 T \max + 0.4 T \min$

Temperatures received at other times of the day are again in the form of 12 two hourly values giving the value of $T_\Delta$ as

$$T_\Delta = \frac{0.75(T_{13} + T_{14} + \cdots + T_{20})}{8} + \frac{0.25(T_{21} + \cdots + T_{24})}{4}$$

where $T_{13} \to T_{24}$ are the two hourly forecast temperatures for tomorrow.

The form of the forecasting equation used is:

$$Y_0 = Y_{-1} + b + a\, \Delta T_H$$

where $Y_0$ = forecast send-out for tomorrow
$Y_{-1}$ = estimate send-out for today
$\Delta T_H$ = change in forecast effective temperature between tomorrow and today.
$a$ and $b$ = regression coefficients.

It must be stressed that these short-term forecasting models only act as a guide to Grid staff who also employ experience and other forecast weather parameters to amend the model result.

## CONCLUSION

The gas industry has found demand forecasting and planning to be much easier and more accurate as a result of using the presently available weather information and services. Meteorological expertise has also been employed from time to time in other ways. For example, the Glasgow Met Office has aided British Gas on several occasions to trace the possible source of gas releases into the atmosphere. It is felt, however, that there is still scope for further accuracy in the weather forecasts which are provided and the industry awaits the improvements that future technology and monitoring will bring.

# Tourism and the Scottish weather

## BRIAN HAY

*Scottish Tourist Board, Edinburgh*

## INTRODUCTION

TOURISM is one of the fastest growing industries in Scotland, with over 200,000 people currently employed in tourist-related activities. It is one of several service industries which are gradually replacing the older manufacturing base. As Scotland moves towards the post-industrial society, the importance of tourism for the economy will become more apparent. Thus, in 1987, tourism trips in Scotland increased by 15% over the previous year (Scottish Tourist Board, 1987a), a growth rate higher than in many others countries. In 1987, nearly 15 million tourist trips were undertaken in Scotland, involving a total of 73 million bednights and an expenditure of almost £2·0 billion.

What is significant, apart from the size of the industry, is that 90% of all tourist trips to Scotland are undertaken by British residents, with over half of such trips involving Scots. These UK domestic tourists are more likely to be aware of the weather in Scotland than overseas visitors, and to plan their activities accordingly. It is also important to note that one-third of all domestic tourism trips, and one in ten of all trips originating overseas, are made by people on business. To a large extent, such trips are not weather dependent. Similarly, about 15% of all domestic tourism trips, and 22% of all overseas-based trips, are made to visit friends and relatives. These trips are also not likely to be too weather sensitive although, if the weather is not considered suitable at the time planned for a visit, the trip may be delayed. Therefore, excluding business trips and visits to friends and relatives, about 50% of all domestic, and 60% of all overseas, trips tend to be weather dependent to some extent. In all, about 7·5 million of these holiday trips took place in Scotland in 1987.

It should also be appreciated that the type of holiday undertaken in Scotland has changed over the past thirty years. The traditional week at a boarding house by the seaside has gone. Today, tourists tend to go abroad for their main holiday if they are looking for guaranteed summer sunshine. The mainstay of the holiday market for domestic tourists is the second holiday and/or the additional holiday, which now accounts for some 60% of all holiday trips. However, over half of all holiday trips still occur in the June-September period. This is not surprising as schools are on holiday at this time and most workers in cities such as Glasgow and Edinburgh have their traditional holiday during the respective Fair and Trades fortnight.

## RELATIONSHIPS BETWEEN TOURISM AND WEATHER

In the first instance, it is necessary to understand how important a deterrent the weather is to holidaymakers. A survey in 1981 sought to identify various factors that

TABLE 17.1. *Unattractive features about Scotland mentioned by tourists*

|  | % of Holidaymakers | |
| --- | --- | --- |
|  | British visitors | Overseas visitors |
| Weather | 11% | 18% |
| Price of Food/Drink | 6% | 8% |
| Price of Petrol/Transport | 6% | 4% |
| Dirty Toilets | 3% | 3% |
| Poor Roads | 5% | 6% |

Source: *STB Scottish Leisure Survey,* 1981.

people found unattractice about Scotland and, as shown in Table 17.1, the weather emerged as the most important feature. Of course, for some holidays 'bad' weather is a positive attribute. It is a truism, for example, to say that without snow there would be no ski-ing. Ski-ing activity has increased markedly over the past decade, with over 700,000 skier days recorded in 1987/88. In 1988, because of the relative lack of snow, the number of ski-ing passes issued declined by 10–15%, with severe economic consequences for local trade in the resort areas.

The traditional Scottish scenes, as presented in the tourist brochures, include the usual mixture of sunny photographs but there is never a picture of rain. Guide books tend to suggest that the east coast of Scotland is relatively dry and sunny. Resorts such as Dunbar, St Andrews, Arbroath and Montrose can boast some of the highest sunshine durations in Scotland with an average, over the summer months, of around 150–200 hours per month (5·6 hours per day). Along the Clyde coast and Moray Firth values are, in general, about 30 minutes less each day. However, resorts along the south coast of England have about two hours of sunshine per day more than is recorded on the east coast of Scotland.

It is difficult to assess the direct effects of weather on tourism. Occupancy rates at camping and touring caravan sites might be considered one of the more weather sensitive indices of holiday accommodation, although it could also be argued that those who undertake such holidays are hardy enough not to be deterred by adverse conditions. Table 17.2 offers some evidence that occupany rates for camping and caravan sites in the Aberdeen area may have been depressed by rain in 1985, compared with the following summer, whilst Table 17.3 also hints at an inverse relationship between the number of domestic tourists and summer rainfall in Edinburgh. July and August 1985 were exceptionally wet in Edinburgh when it might be expected that tourist trips would be around 15–18% rather than the 8–9% actually recorded.

Another measure of weather sensitivity is obtained by looking at the activities people engage in when on a day trip. During 1987 a survey of leisure day trips was carried out and some preliminary results of participation rates in various outdoor activities are presented in Table 17.4 (Scottish Tourist Board, 1987b). Although

TABLE 17.2. *Occupancy rates at camping and caravan sites in Aberdeen in 1985 and 1986 and rainfall*

|  | 1985 | | 1986 | |
| --- | --- | --- | --- | --- |
|  | Occupancy rates | Monthly rainfall | Occupancy rates | Monthly rainfall |
| April | 7% | 76 mm | 6% | 80 mm |
| May | 13% | 95 mm | 16% | 84 mm |
| June | 31% | 116 mm | 28% | 60 mm |
| July | 51% | 136 mm | 56% | 51 mm |
| August | 46% | 146 mm | 64% | 104 mm |
| September | 12% | 100 mm | 14% | 21 mm |
| October | 7% | 9 mm | 9% | 45 mm |

Source: *STB Occupancy Studies 1985, 1986*;

concerned solely with day trips by Scottish residents, it confirms that people increase their participation rates in outdoor activities during the summer. A similar seasonal trend exists in relation to the location of day trips where it is known that, in the summer months, the popularity of countryside and seaside venues increases relative to visits to towns and cities. On a more detailed time-scale, Fig. 17.1 shows the weekly number of visitors to Strathclyde Regional park compared with rainfall totals and mean maximum temperatures. Visitor numbers and temperatures tend to follow a broad seasonal pattern but there appears to be some evidence for a short-term,

TABLE 17.3. *Monthly distribution of tourists and rain-days in Edinburgh in 1985*

|  | % of tourists | Days of rain |
| --- | --- | --- |
| January | 3% | 15 |
| February | 5% | 6 |
| March | 7% | 16 |
| April | 7% | 17 |
| May | 6% | 13 |
| June | 15% | 13 |
| July | 8% | 18 |
| August | 9% | 25 |
| Sept | 20% | 17 |
| October | 6% | 11 |
| November | 6% | 15 |
| December | 7% | 21 |

Source: *STB National Survey of Tourism in Scotland 1985*.

*Tourism and the Scottish weather*

TABLE 17.4. *Rates of participation in outdoor activities on leisure day trips in Scotland*

|  | Month of trip | | | | | |
|---|---|---|---|---|---|---|
| Activity undertaken | Feb | Apr | June | July | Aug | Oct |
| Gone on a Tour/Drive | 11% | 23% | 22% | 29% | 36% | 15% |
| Gone on a Long Walk/Hike/Ramble | 12% | 15% | 15% | 17% | 19% | 9% |
| Visited a Park | 6% | 15% | 13% | 17% | 18% | 8% |
| Visited a Historic/Ancient Site/Monument | 3% | 7% | 11% | 15% | 17% | 8% |
| Gone Cycling | 3% | 3% | 2% | 7% | 2% | 3% |

Source: *STB Leisure Day Trip Survey*, 1987.

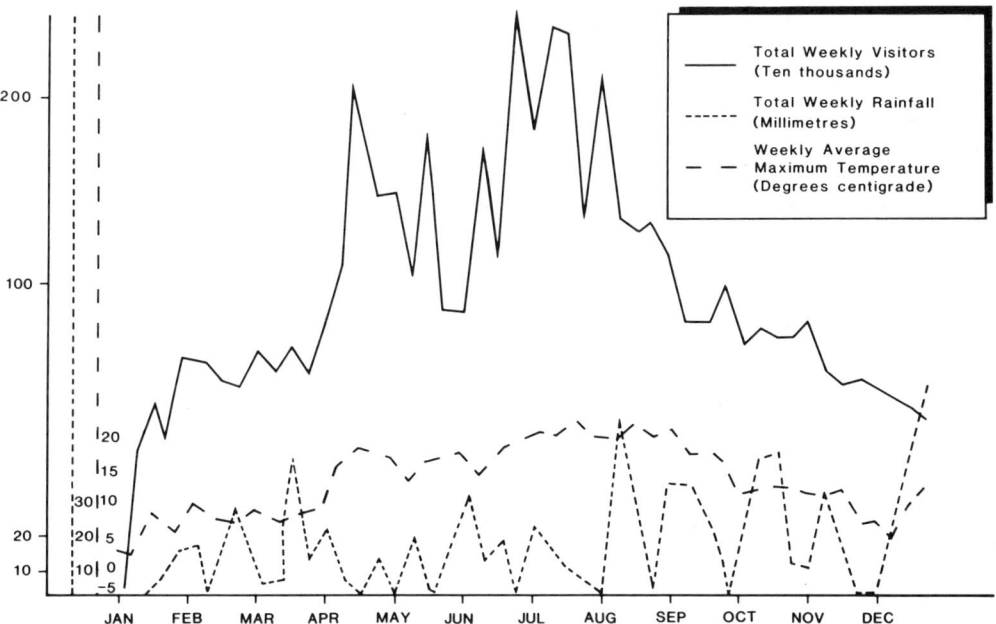

FIG. 17.1. Weekly visitors to Strathclyde Regional Park in 1987 compared with rainfall and temperature conditions.

inverse relationship between attendance and rainfall, although a specific study would be required to demonstrate this properly.

## SCOTTISH TOURIST BOARD'S REACTION TO THE WEATHER

In Scotland the late spring and early summer period often produces the driest and sunniest weather of the year. Average monthly rainfall would certainly appear to favour the early holidaymaker. For example, in Edinburgh, the 30-year average rainfall is 39 mm in April, 52 mm in May and 45 mm in June whereas in July and August—the conventional holiday period—the values are 69 mm and 79 mm respectively. The advantageous weather in late spring and early summer fits in well with the Tourist Board's policy of promoting holidays and short breaks in the off-season months.

The typical tourists who holiday in Scotland are not looking for fine weather but for activities and attractions. Most visitors use a car and a hot day for driving is not very comfortable. Some activity holidays can also be less pleasant in such conditions. One definite advantage of Scotland's latitudinal position is the long twilight evenings which give more than three hours extra daylight in mid-summer relative to resorts on England's south coast.

The Scottish Tourist Board has recognised the importance of the weather when drawing up development priorities. Grant aid is given for the development of new and improved visitor attractions, enhancement of the existing stock of serviced accommodation and for new and improved recreational facilities. All of these priority initiatives reflect the importance of indoor facilities designed, to some extent, to counter the Scottish weather. The provision of indoor facilities is not new. Indeed, such developments as the Edinburgh and Glasgow Botanic Gardens and Aberdeen's winter gardens, all with large heated greenhouses, are early examples of sheltered recreational space for tourists in inclement weather. Zoos and municipal swimming pools are further examples of this emphasis. The policy continues today as exemplified by the Scottish Tourist Board's recent £1·2 million grant to Butlin's at Ayr to develop indoor recreational facilities.

In conclusion, it should be recognised that, given the variability of Scotland's weather, a sunny holiday can never be guaranteed. But it is suggested that the vast majority of people who holiday in Scotland do not expect such an experience and visit the country for other reasons.

# CONSPECTUS

# Weather sensitivity and the future for meteorological services

## S. J. HARRISON

*Department of Environmental Science, University of Stirling*

### DEFINING WEATHER SENSITIVITY

To the British public the weather is a topic central to many conversations yet there exists no common perception of weather conditions and weather forecasts (Ayton, 1988) and the level of understanding of weather processes is generally low. There is, nevertheless, evidence of a growing awareness, particularly amongst the business and public service communities, that the weather imposes measurable costs which may be reduced through the input of appropriate meteorological data into decision-making processes. The central aim of the symposium which led to this volume was to encourage this trend and to examine ways in which meaningful dialogues can be developed between the purveyor of meteorological services and potential customers. In order to move towards this objective, it was necessary to identify the nature of the meteorological products in terms of hardware, software and services, and the specific needs of the customer, and ultimately to agree on an agenda for future discussions.

The contributors to the symposium have established beyond doubt that the range of weather impacts on economic activities is wide. These are not of the catastrophic magnitude of, for example, the tropical cyclone, but do include discrete high impact events, the best documented of which are winter snowstorms, which affect many of Scotlands roads (Brinham: Chapter 6) and the Scotrail system (Crawford: Chapter 7), and floods (Sargent: Chapter 8). However, a significant proportion of costs directly attributable to the weather probably arises not as a result of high magnitude low frequency events but rather through the persistence of low magnitude events. For example, a particular characteristic of the Scottish climate is the frequency with which long spells of rain occur within which intensities are relatively low. The most recent example was the summer of 1985 (Smith: Chapter 1) during which the longest rain-free spell in many places was only three days, resulting in considerable losses in agricultural production. Prior (Chapter 14) suggests that in the context of the building and construction industries persistence, rather than quantity, of rainfall is a more meaningful measure of adversity of weather conditions.

The question remains, however, as to whether sensitivity has been, or indeed could be, adequately defined as it clearly implies a graduated scale of quantifiable impact. To achieve this is a tall order given the complex nature of weather events and the wide perceptual spectrum of responses to them, together with the difficulty of attributing specific costs. These difficulties are well illustrated in the contracts relating to construction work where the terms 'adverse' and 'exceptional' are used in legal documents without any satisfactorily objective method of defining either (Carson: Chapter 13). Some bodies, such as British Gas, have, however, managed to define a sliding scale of temperature impact on the supply of gas to customers. In most cases the interpretation of weather impact will be defined for specific areas of activity and will contain a significant element of subjectivity. Setting this aside, there remain a number of more tangible problems which must be addressed before entering into any discussion of the relative merits of different marketing strategies for meteorological services.

## DEFINING THE PROBLEMS

One of the main difficulties in trying to assess weather character in Scotland is that of achieving adequate precision in description and forecasting in a mid-latitude maritime setting characterised by small-scale spatial and temporal variability and in which there is a far from adequate database. The general features of the climate are determined by Atlantic weather systems which are dominant for most of the year and which bring a high expectation of precipitation. For many activities, such as agriculture, the Scottish climate has been referred to as being 'hostile' (Callander: Chapter 9), while for others it imposes a seasonality in demand for services such as the salting of roads, or in environmental conditions affecting specific activities such as the construction industries, where there are operational thresholds.

The diverse nature of surface topography adds further complexity. Slope, aspect, principal orientations of topographic features with respect to airflow and the movement of major weather systems, and elevation combine to produce climatic variations over small spatial scales. The effects of surface elevation, in reducing solar radiation and temperature and increasing wind speed and precipitation, have not been accurately quantified and much use is made of crude and usually untested linear climate-elevation models (Harrison and Harrison, 1988). Poor access and remoteness, together with a harsh climatic environment, are obvious deterrents to weather observation and have tended to discourage weather sensitive activities in what are frequently referred to as marginal environments (Jenkins *et al.*, 1974). A basic lack of information has obvious implications in a country where principal lines of communication, whether these be for wheeled transport or in the form of overhead transmission lines, are frequently more than 300 m above sea-level. The season of risk from severe weather conditions associated with ice and snow accumulation is considerably longer at these vulnerable points within communication systems. The expansion of forestry and wind power industries into the Scottish uplands adds urgency to improving the database for these areas for which the current mesoscale numerical forecasts issued by the Met Office have insufficient spatial resolution.

The Met Office receives a large amount of data from stations throughout Scotland, the most numerous of which are the climatological stations which have a mean spatial representation of 1 per 453 km$^2$. The distribution, as previously indicated, is far from uniform and there are major gaps in the more remote upland areas. One of the implications is that knowledge of current weather and the progress of weather systems will be incomplete. Reynolds (1985), for example, refers to a case in north-west Scotland where localised precipitation was almost completely missed by the rainfall observing network. In other locations stations may not record at time intervals most suited to operational decision making, being most commonly in the form of 24 hour (0900–0900 GMT) extremes and totals. A large number of stations also do not have a sufficient length of record to support the calculation of statistical probabilities of exceedance for defined operational thresholds.

Apart from these obvious deficiencies a conventional surface observation network will, by its very nature, be insufficiently responsive to changes in the demand for data, in terms of location, parameters measured, and temporal resolution of sampling. What was appropriate to local health authorities in the nineteenth century is unlikely to be best suited to the needs of high technology industries in the 1990s. There is, therefore, a real need to explore ways of improving the supply of data to an ever changing market and perhaps to move away from conventional observation practices and an archive mentality. The observing network could more usefully contain both fixed, long-term, reference stations and mobile short-term stations deployed for periods of the order of a year. The climatological stations will, however, continue to have value in the formulation of statements of long-term risks.

In an age where everything seems to be required by yesterday, time is clearly an important consideration in meeting the demands for meteorological services. The dual problems are the timescale of the forecast and the rapidity with which it can be communicated. There is a tendency for the accuracy of meteorological prognoses to increase as timescales decrease, culminating in the description of current weather, or the 'nowcast' (Browning, 1982). Probabilistic statements, although not strictly a forecast, represent the longest timescale and are incorporated at project planning and design stages. The problems here are not of communication time but of the adequacy of databases. Of the meteorological, as distinct from climatological, forecasts potentially the most saleable is the long-range forecast for more than a month ahead. Early attempts at long-range forecasting, based on analogue methods, were abandoned by the Met Office as accuracy was low, but new modelling techniques are prompting a re-appraisal.

Shorter range forecasting up to 24 hours ahead is the most valuable for operational decision-making. The forecast at this timescale is at its most reliable but the demands are for rapid communication and updating, and tailoring to often very specific operational needs. The implication is that the forecaster needs to be acquainted with the customer's work practices and is more likely to require site-specific and operationally orientated data. For example, the Met Office's 'Open Road' service, which provides a comprehensive 24 hour forecast of road surface conditions together with regular updates, has required the introduction of observation systems for the continuous monitoring of road surface conditions (Brinham: Chapter 6). There is also the prospect of the introduction of on-site automatic weather stations to assist in construction operations. These raise the important problem of matching monitoring standards as the forecast, however reliable its meteorological basis, will clearly be vulnerable to indifferent site-specific data quality. For example, it has been generally agreed that, in the siting of road surface sensor equipment, more attention could have been paid to meteorological criteria.

Those who have invested in meteorological services appear to have achieved measurable reductions in operational costs (Brinham: Chapter 6). However, the fact remains that these represent an enlightened minority in Scotland. Callander (Chapter 9), for example, has referred to the very poor uptake of forecast services by farmers in an industry which, on the basis of past experience, is clearly very vulnerable to the vagaries of the Scottish climate. Questionnaire-based research into the reasons for this failure to attract attention to meteorological services is clearly required but at this stage it would appear that it could be attributed to:

(a) a general lack of perception of weather-related losses as being avoidable and a basic lack of understanding of weather mechanisms.

(b) a reputation which has been based on forecasts to the general public which may be perceived as being unreliable, particularly in terms of timing, and insufficiently precise.

(c) an expectation on the part of the customer that meteorological data appropriate to any particular location should be available and that extrapolations from distant stations are unreliable in a country where local climatic differences are readily apparent.

(d) the fragmented nature of the market for meteorological services. It is difficult to develop a forecasting service which is applicable across the whole width of a particular industrial sector. It is no coincidence that dedicated forecasting for roads departments has been amongst the first to take off as a basic commonality of problem can be identified and addressed irrespective of geographical location.

(e) the under-marketing of meteorological services and a tendency to allow the market to be led by the customer. The arrival of private forecasting agencies, such as Noble Denton and Weather Watchers, has amply demonstrated the rewards of more aggressive marketing policies.

Faced with these problems and the emergence of competition, the question must be asked as to how national meteorological agencies, such as the Met Office, move ahead into a new and considerably busier market place. There is a clear need to develop and improve the basic product while at the same time developing marketing strategies which are to the mutual benefit of both supplier and customer.

## THE IMMEDIATE PROSPECTS

While forecasting reliability continues to improve, and remotely sensed information becomes more generally available, there is a great deal that remains to be achieved in the marketing of meteorological products. In developing a strategy there are a number of immediate questions which must be addressed if the quality and usefulness of the products is to be maintained and if their introduction into decision-making is to have a sound long-term future.

It is clear that the market requires more flexible observation systems better capable of responding to changing needs than the existing Met Office climatological network, which conforms in most respects to laid down standards. There are alternative sources of data available but care must be taken to ensure that adequate levels of quality control are maintained (Pettifer: Chapter 3) and that multiple, and possibly mutually independent, standards do not emerge. There is a growing body of observers in organisations such as the Climatological Observers Link (Rolfe, 1982) and the Register of Weather Stations (Harrison, 1982) whose observation practices in many cases fall short of the standards laid down by the Met Office. While all are enthusiastic and diligent, it must be remembered that observations made with, for example, a Six's thermometer on the north side of a wooden post can not readily be compared with those from a conventional thermometer screen. There is no reliable way of translating such non-standard observations into a standard form yet there is real danger that data from such sources will be increasingly found in the market place. The informed amateur can, however, provide very useful observations of a more visual nature (cloud cover and snow lie for example) and of a binary form (is it raining or not), as demonstrated by the Weather Watchers Network in Scotland. Such a low technology data source, while being potentially expensive to collate, could usefully complement currently available short-term weather forecasts.

At the other end of the scale there is the question of the future role of automated weather data retrieval which in itself generates a number of subsidiary questions pertaining to quality control, customer use, and integration with conventional observations. The systems range from the automatic weather station (AWS), such as the Harsh Environment AWS being developed by the Met Office, to satellites and the three radar installations promised for Scotland, the first of which is due to be in place in the central lowlands by the summer of 1990. The AWS offers opportunities to fill gaps in the existing observing network, increases flexibility, and can improve the temporal resolution of the data gathered. However, correct deployment, particularly in terms of siting, is critical otherwise their widespread use may pose problems where interfacing of the data with weather forecasts is required. The FRONTIERS system described by Collier (Chapter 2) merges remotely sensed data from satellite, radar and surface stations in producing descriptions of current weather patterns and forecasts over the following few hours. The introduction of such a system to Scotland alongside promised improvements in the resolution of numerical forecasts to 1–2 km appears to offer what the market requires for operation decision-making and may well form the basis for a range of specialist forecasts in addition to the enhancement of existing forecast services such as 'Open Road'.

## SETTING OUT THE STALL

The important question remains as to whether potential customers will see improved meteorological services as being of value and worthy of purchase. Certainly there would be little point in improving forecasts without reference to the specific needs of customers or without exercising care in their packaging. Effort also needs to be directed into heightening general awareness of the potential value of weather information. The market place for weather services is currently undergoing very rapid change as some private forecasting agencies, anxious to secure their own profitability, engage in more determined marketing. If competition results in a better service to the customer, and a better public image of meteorological services, then this is all to the good but it is vital that agencies exist to maintain standards. The meteorological profession would surely not wish to see litigation as the basis on which the market place is purged of poor practices.

As far as the customer is concerned, there is obviously a great deal that needs to be done to ensure that forecasts, and nowcasts are targeted at the most beneficial areas of operation and that supplementary meteorological observation is carried out in the most appropriate manner. A potentially useful way forward would be to organise workshops on specific aspects of the supplier-customer interface. Their principal functions would be to initiate dialogue, to identify the principal needs of the customer, and to examine how best these can be met. A programme of workshops has been planned by the Climatic Hazards Unit of the University of Stirling, the first of which is directed at tourism and outdoor recreation in Scotland.

The symposium represents the first step along what may be a long and, at times, difficult road. No doubt political and economic considerations will play a large part in shaping the future for meteorological services and their role in society but it is hoped that at the end of the day the customer and, in particular, the Scottish economy will be major beneficiaries. Meteorological services are certainly at a very important crossroads, as indicated by Smith (1988b). It is vital that the correct decisions are made over the next few years and that agencies work together for the common good. There is a very clear role for a central agency able to coordinate practices and to ensure that supply and demand are, as far as is possible, matched.

# References

ABRAHAMSON, M. W. (1979) *Engineering Law and the ICE Contracts*. 4th Edition, Elsevier Applied Science, London.

ALLAM, R. (1987) The detection of fog from satellites. Preprints *Workshop on Satellite and Radar Interpretation*, Reading, England, 20-24 July, EUMETSAT, pp 495-507.

ANONYMOUS (1987) *Regional Climate Centers: A New Institution for Climate Services in the United States*. An unpublished document prepared by staff of the regional climate centers and the National Climate Program Office, Washington, D.C. 12pp.

APSIMON, H. M., SIMMS, K. L. & COLLIER, C. G. (1988) The use of weather radar in assessing deposition of radioactivity from Chernobyl across England and Wales. *Atmospheric Environment* 22 (9), 1895-1900.

ATTERSON, J. (1980) *Gambling with Gales*. An unpublished paper presented to the British Association for the Advancement of Science, Section K (Forestry), Salford.

AYTON, P. (1988) Perceptions of broadcast weather forecasts. *Weather* 43, 193-197.

BAILEY, J. D., BARRETT, E. C., HARRISON, A. R., HERSCHY, R. W., LUCAS, R. M. & POWER, C. H. (1987) *Satellite Remote Sensing in Hydrology and Water Management*. Final report to DOE, June, 273 pp.

BEESLEY, M. and BUDD, A. (1987) *An Economist's Overview—Costs, Benefits and Penalties*. An unpublished paper presented in the Introductory Notes to the seminar on *Winter Chaos*, Institution of Civil Engineers, London, pp 29-35.

BELL, R. S. & DICKINSON, A. (1987) The Meteorological Office operational numerical weather prediction system. *Met Office Scientific Paper* No. 41, H.M.S.O. 61 pp.

BENGTSSON, L. (1985) Medium-range forecasting at the ECMF. In MANABE, S. (Ed) *Advances in Geophysics* 28B, Academic Press, New York and London.

BERTNESS, J. (1980) Rain-related impacts on selected transportation activities and utility services in the Chicago area. *Journal of Applied Meteorology* 19, 545-556.

BOOTH, B. J. (1984) Applications of automated weather radar and Meteosat displays in an aviation forecast office. *Meteorological Magazine* 113, 32-42.

BROWN, P. R. (1961) Rain and/or low temperature as factors interrupting external building work in the Glasgow area. *Climatological Memorandum* No. 30, Met Office, Bracknell.

BROWNING, K. A. (1978) Meteorological aspects of radar. *Reports of Progress in Physics* 41, 761-806.

BROWNING, K. A. (1979) The Frontiers plan: a strategy for using radar and satellite imagery for very short-range precipitation forecasts. *Meteorological Magazine* 108, 161-183.

BROWNING, K. A. (1981) A total system approach to a weather radar network. *Proc. IAMP Symposium*, Hamburg, 25-28 August, ESA SP-165, pp 115-122.

BROWNING, K. A. (1982a) Exremely low temperatures over England and Wales observed by Meteosat 2. *Weather* 37, p 79.

BROWNING, K. A. (1982b) (Ed) *Nowcasting* Academic Press, London.

BROWNING, K. A., COLLIER, C. G., LARKE, P. R., MENMUIR, P., MONK, G. A. & OWENS, R. G. (1982) On the forecasting of frontal rain using a weather radar network. *Monthly Weather Review* 110, 534-552.

BRUGGE, R. (1987) Low daytime temperatures over England and Wales on 12 January 1987. *Weather* 42, 146-152.

BURRIDGE, D. M. (1979) Some aspects of large scale numerical modelling of the atmosphere. *Proc. of ECMWF Seminar on Dynamical Meteorology and Numerical Weather Prediction* Vol 2, pp 1-78.

BURRIDGE, D. M. & GADD, A. J. (1977) The Meteorological Office operational 10-level numerical weather prediction model (December 1975). *Meteorological Office Scientific Paper* No. 34, HMSO, London.

BUSHBY, F. H. (1987) A history of numerical

weather prediction. In MATSUNO, (Ed) *Short and Medium-Range Numerical Weather Prediction,* Meteorological Society of Japan, Tokyo.

CARPENTER, K. M. (1979) An experimental forecast using a non-hydrostatic mesoscale model. *Quarterly Journal Royal Meteorological Society* **105,** 629–655.

CARPENTER, K. M. & BROWNING, K. A. (1984) FRONTIERS—progress with a system for nowcasting rain. *Proc. Nowcasting II,* Norrkoping, 3–7 September, ESA SP-208, pp 427–432.

CHEN, R. S. & PARRY, M. L. (1987) *Policy-orientated Impact Assessment of Climatic Variations.* International Institute for Applied Systems Analysis, Laxenburg, Austria. 54 pp.

COLLIER, C. G. (1986) Accuracy of rainfall estimates by radar, Part 1: Calibration by telemetering raingauges. *Journal of Hydrology* **83,** 207–223.

COLLIER, C. G. & LARKE, P. R. (1978) A case study of measurement of snowfall by radar: an assessment of accuracy. *Quarterly Journal Royal Meteorological Society* **104,** 615–621.

COLLIER, C. G. & JAMES, P. K. (1986) On the development of an integrated weather radar data processing system. *Preprints 23rd Conference Radar Meteorology* 22–26 September, Snowmass, Colorado. American Meteorological Society, Boston, JP95–98.

COLLIER, C. G., LARKE, P. R. & MAY, B. R. (1983) A correction procedure that can be applied to weather radar measurements to allow real-time estimation of surface rainfall. *Quarterly Journal Royal Meteorological Society* **109,** 589–608.

COLLIER, C. G., FAIR, C. A. & NEWSOME, D. H. (1988) International weather radar networking in Western Europe. *Bulletin American Meteorological Society* **69** (1), 16–21.

COOK, N. J., SMITH, B. W. & HUBAND, M. V. (1985) BRE Program Strongblow. *The Designers Guide to Wind Loading of Building Structures* (Supplement Two), Building Research Establishment, Garston.

DEFREITAS, C. R. (1975) Estimation of the disruptive impact of snowfalls in urban areas. *Journal of Applied Meteorology* **14,** 1166–73.

DIALLO, N. T. & FRANK, W. M. (1986) Effects of enhanced initial moisture fields on simulated rainfall over West Africa and the East Atlantic. *Monthly Weather Review* **114,** 1811–1821.

EDMOND, G. (1985) Recent events and developments in winter maintenance of roads in the Highland Region. In HARRISON S. J. (Ed.) *Climatic Hazards in Scotland* Geo Books, Norwich, pp 67–71.

EYRE, J. R. & JERRET, D. (1982) Local-area atmosphere sounding from satellites. *Weather* **37,** 314–322.

EYRE, J. R., BROWNESCOMBE, J. L. & ALLAM, R. J. (1984) Detection of fog at night using an Advanced Very High Resolution Radiometer (AVHRR) *Meteorological Magazine* **113,** 266–271.

FINDLATER, J., ROACH, W. T. & MCHUGH, B. C. (1989) The haar of north-east Scotland. *Quarterly Journal Royal Meteorological Society* (in press).

FERGUSON, H. L. & PHILLIPS, D. W. (1986) *Developing a National Climate Program— A Decade of Canadian Experience.* Canadian Climate Program, Downsview, Ontario, 30 pp.

FLATHER, R. A. (1981) Practical surge prediction using numerical models. In PERIGRINE, (Ed) *Floods due to High Winds and Tides.*

FLOOD, C. R. (1985) Forecast evaluation. *Meteorological Magazine* **114,** 254–260.

FOLLAND, C. K. & WOODCOCK, A. (1986) Experimental monthly long-range forecasts for the United Kingdom. Part 1. Description of the forecasting system. *Meteorological Magazine* **115,** 302–318.

FORTH RIVER PURIFICATION BOARD (1984) *Annual Report for 1984,* Forth RPB, Colinton, Edinburgh.

FRANCIS, P. E. (1981) The climate of the agricultural areas of Scotland. *Climatological Memorandum No. 108,* Met Office, Bracknell.

GADD, A. J. (1985) The 15-level weather prediction model. *Meteorological Magazine* **114,** 222–226.

GLOYNE, R. W. (1968) Some climate in-

## References

fluences affecting hill land productivity. *Symposim on Hill Land Productivity*, European Grassland Federation, July.

GLOYNE, R. W. (1971) Notes and some speculations on the effects of climate on the development of horticulture in Scotland. *Scientific Horticulture* **23**, 22–40.

GOLDING, B. W. (1983) A wave prediction system for real-time sea state forecasting. *Quarterly Journal Royal Meteorological Society* **109**, 393–416.

GOLDING, B. (1987a) Strategies for using mesoscale data in an operational mesoscale model. *Preprints Workshop on Satellite and Radar Imagery Interpretation* Reading, England, 20–24 July, EUMETSAT, pp 341–364.

GOLDING, B. W. (1987b) Short-range forecasting over the United Kingdom using a mesoscale forecasting system. In MATSUNO, (Ed) *Short and Medium-Range Numerical Weather Prediction*, Meteorological Society of Japan, Tokyo.

GOLDING, B. W. (1988) The use of numerical models in weather forecasting-achievements and prospects. In HARRISON, S. J. and SMITH, K. (Eds) *Weather Sensitivity and Services in Scotland*, Scottish Academic Press, pp. 40–54.

HARPER, W. G. (1974) Irrigation needs in the main agricultural regions of Scotland. *Agricultural Memorandum No. 634*, Met Office, Bracknell.

HARRISON, S. J. (1980) Rainfall in the Stirling area. *Forth Naturalist and Historian* **5**, 23–24.

HARRISON, S. J. (1982) The ROWS survey of stations. *Weather* **37**, 82–83.

HARRISON, S. J. & HARRISON, D. J. (1988) The effect of elevation on the climatically determined growing season in the Ochil Hills, Scotland. *Scottish Geographical Magazine* **104** (2), 108–115.

HARVERSON, D. (1985) Ice warning systems on British roads. *Highways* **53**, 26–27.

HMSO (1981) *Handbook of Meteorological Instruments* Vol. 4. Met. O 919d. Her Majesty's Stationery Office, London.

HOUGHTON, D. M. (1987) *Responses to Climatic Variability* An unpublished paper presented in typescript to the workshop on *The Implications of Climatic Variability for Industry in the United Kingdom*. University of Birmingham, 6 pp.

HUNT, R. D. (1987) *The Meteorological Office View*. An unpublished paper presented in the Introductory Notes to the seminar on *Winter Chaos*, Institution of Civil Engineers, London, pp 1–5.

HUNTER, R. H. & COLLINS, A. (1987) Initial estimates and measurements of the wind climate at the National Wind Turbine Centre's test site, Myres Hill.

JARAS, T. F. (1987) *Wind Energy 1987— Wind Turbine Shipments and Applications*. Wind Data Center, Stadia Inc., Virginia, U.S.A.

JATILA, E. (1973) Experimental study of the measurement of snowfall by radar. University of Helsinki, Dept. of Meteorology Paper No. 122, *Geophysica* **12**, 1–10.

JENKINS, D. A., OXLEY, E. R. B. & GETHING, P. A. (1974) *Marginal Land: Integration or Competition*. Fourth Colloquium of the Potassium Institute, Henley.

KEEBLE, E. J. & PRIOR, M. J. (1988) *Climate and Construction Operations in the Plymouth Area* Building Research Establishment, Garston.

KING, E. G. E. (1981) Daytime raintime. *Site Management Information Service* No. 87, Chartered Institute of Building, Ascot.

LACY, R. E. (1977) *Climate and Building in Britain*. Building Research Establishment, Garston.

LAMB, P. J. (1981) Do we know what we should be trying to forecast-climatically? *Bulletin of the American Meteorological Society* **62**, 1000–1.

LAMB, P. J., SONKA, S. T. & CHANGNON, S. A. (1985) *Use of Climate Information by U.S. Agribusiness*. NOAA Technical Report NCPO 001, National Climate Program Office, Rockville, Maryland, 67 pp.

LEARMONTH, A. T. A. (1950) The floods of 12 August 1948 in south-east Scotland. *Scottish Geographical Magazine* **66**, 147–153.

LEE, A. C. L. (1981) Smoothing and filtering of meteorological data. *Meteorological Magazine* **110**, 115–132.

LLEWELLYN-JONES, D. T., MINNETT, P. J.,

# References

Saunders, R. W. & Zavody, A. M. (1984) Satellite multichannel infra-red measurements of sea surface temperature of the N.E. Atlantic Ocean using AVHRR/2. *Quarterly Journal Royal Meteorological Society* **110,** 613–631.

Manley, G. (1970) The climate of the British Isles. In *Climates of Northern and Western Europe, World Survey of Climatology Vol 5*, Elsevier, Amsterdam.

Marsh, T. & Lees, M. (1985) *The 1984 drought.* Institute of Hydrology, Wallingford.

May, B. R. (1988) Progress in the development of PARAGON. *Meteorological Magazine* **117,** 79–86.

McAlonan, W. S. (1984) *Winter Maintenance.* An unpublished paper presented to a meeting of the Institution of Highways and Transportation, Galashiels, 8 pp.

Maunder, W. J. (1986) *The Uncertainty Business.* Methuen and Co. Ltd., London, 420 pp.

Meteorological Office (1971) The climate of Scotland—a brief review. *Select Committee on Scottish Affairs*: Session 1971–72 V, Appendix A36, pp 175–187.

Meteorological Office (1985) *Services to the Construction Industry.* Met Office, Bracknell.

Monk, G. (1987) Topographically related convection over the British Isles. In Bader, and Waters, (Eds) *Satellite and Radar Imagery Interpretation.*

National Oceanic & Atmospheric Administration (1980) *National Climate Program*: *Five-Year Plan*. N.O.A.A., Washington, D.C., 123 pp.

Palutikof, J. P., Davies, T. D., Kelly, P. M. & Halliday, J. A. (1986) Temporal and spatial variations in hourly windspeeds over the British Isles. EWEA.

Parker, D. E. & Folland, C. K. (1987) *The Nature of Climatic Variability.* An unpublished paper presented in typescript to the workshop on the *Implications of Climatic Variability for Industry in the United Kingdom.* University of Birmingham, 30 pp.

Parker, K. T., Lord, W. B. H., Read, N. J. & Parsons, J. (1986) *Social and Economic Responses to Climatic Variability in the UK.* Technical Change Centre, London, 160 pp.

Penning-Rowsell, E. C. & Chatterton, J. B. (1977) *The Benefits of Flood Alleviation.* Saxon House, Teakfield, Ltd.

Perry, A. H. & Symons, L. (1980) The economic and social disruption arising from the snowfall hazard in Scotland—the example of January 1978. *Scottish Geographical Magazine* **96,** 20–25.

Perry, A. H., Symons, L. & Williams, P. J. (1984) Snow depth and snowfall disruption in Scotland in January 1984. *Journal of Meteorology* **9,** 133–135.

Petersen, E. L. & Troen, I. (1988) *European Wind Atlas* Published for the Commission of the European Communities by Riso National Laboratory, Denmark. (In press).

Pettifer, R. E. W. (1981) Automatic meteorological observations: new methods and new problems. *Instruments and Methods of Observation Report No. 9* World Meteorological Organisation, Geneva, 328 pp.

Pettifer, R. E. W. (1984) Automatic systems in operations use—quality and reliability. *Instruments and Methods of Observation Report No. 15* World Meteorological Organisation, Geneva, pp 339–343.

Pick, D. R. (1986) The operational sounding of the low atmosphere from satellites using millimetre waves. *Proc. 16th European Microwave Conference,* Dublin, 8–12 September.

Pielke, R. A. (1981) Mesoscale dynamic modelling. *Advances in Geophysics* **23,** 186–344.

Pollock, D. M. & Wilson, J. W. (1972) Basin precipitation—land and lake. *IFYGL Technical Plan* Vol. 1, pp 107–112.

Ponting, J. F. & Sarson, M. A. (1984) Operational quality evaluation of surface observations. *Instruments and Methods of Observation Report No. 15* World Meteorological Organisation, Geneva, pp 239–244.

Prior, M. J. & King, E. G. E. (1981) Weather forecasting for construction sites. *Meteorological Magazine* **110,** 260–266.

Reynolds, G. (1985) Extreme rainfall

## References

events in Scotland. In HARRISON S. J. (Ed) *Climatic Hazards in Scotland.* Geo Books, Norwich, pp 15–23.

ROBINS, N. S. (1987) Development of groundwater resources in Scotland. *Proceedings Institution of Civil Engineers* Part 2, **83**, 747–753.

ROESLI, H. P., JOSS, J. & COLLIER, C. G. (1987) COST-73 and its application in very short-range forecasting. *Proc. Symposium on Mesoscale Analysis and Forecasting,* Vancouver Canada, 17–19 August, ESA SP-282, pp 13–18.

ROLFE, G. W. (1982) The Climatological Observers Link. *Weather* **37**, 84–85.

ROONEY, J. F. (1967) The urban snow hazard in the United States: appraisal of disruption. *The Geographical Review* **57**, 538–559.

RUSSO, J. A. (1971) *The Complete Money Saving Guide to Weather for Contractors.* Environmental Information Services, Connecticut, USA.

SARGENT, R. J. (1985) *Telemetry for a Flood Prevention Scheme.* Annual Conference, Institute of Water Pollution Control, Bournemouth.

SCOTTISH DEVELOPMENT DEPARTMENT (1973) *Water Supply in Scotland: March 1973.* An unpublished typescript, Edinburgh, 13 pp.

SCOTTISH TOURIST BOARD (1987a) *National Survey of Tourism in Scotland.* Scottish Tourist Board, Edinburgh.

SCOTTISH TOURIST BOARD (1987b) *Scottish Leisure Day Trip Survey.* Scottish Tourist Board, Edinburgh.

SEVRUK, B. (1984) Comparison of evaporation losses from standard precipitation gauges. *Instruments and Methods of Observation Report No. 15* World Meteorological Organisation, Geneva, pp 57–61.

SMITH, C. S. (1986) The reduction of errors caused by brightband in quantitative rainfall measurements made using radar. *Journal Atmospheric and Oceanic Technology* **3**, 129–141.

SMITH, D. H. & RAWLINGS, B. (1974) *The Effect of Adverse Weather on Building Productivity.* Construction Industry Research and Information Association, Report No. 50, CIRIA, London.

SMITH, F. B. (1987) Plan for developing a numerical model capable of simulating the transport, dispersion and deposition of radionuclides released into the earth's atmosphere in real-time. *Unpublished Report,* Met Office.

SMITH, K. (1974) Climatology and hydrology. In TIMMS, D. W. G. (Ed) *The Stirling Region,* University of Stirling, pp. 47–65

SMITH, K. (1977) Water resource management in Scotland. *Scottish Geographical Magazine* **93**, 66–79.

SMITH, K. (1981) The effect of weather conditions on the public demand for meteorological information. *Journal of Climatology* **1**, 381–393.

SMITH, K. (1982a) How seasonal and weather conditions influence road accidents in Glasgow. *Scottish Geographical Magazine* **98**, 103–114.

SMITH, K. (1982b) The influence of snow, fog and heavy rain on the demand for road transport information at Glasgow Weather Centre. *Meteorological Magazine* **111**, 291–296.

SMITH, K. (1983) Weather thresholds for building. *Weather* **38**, 277–281.

SMITH, K. (1985) Climatic hazards in Glasgow: A view from the press. In HARRISON S. J. (Ed.) *Climatic Hazards in Scotland,* Geo Books, Norwich, pp 51–66.

SMITH, K. (1988a) Highway meteorology comes to Scotland. *Scottish Geographical Magazine* **104**, 60–62.

SMITH, K. (1988b) Future trends in atmospheric data and services. *Weather* **43**, 401–405.

SMITH, W. L., WOOLF, H. M., HAYDEN, C. M., WORK, D. Q., MCMILLIN, L. M., TAPP, M. C. & WHITE, P. W. (1979) The TIROS-N operational vertical sounder. *Bulletin American Meteorological Society* **60**, 277–296.

TABONY, R. C. (1985) Relations between minimum temperature and topography in Great Britain. *Journal of Climatology* **5**, 503–520.

TAPP, M. C. & WHITE, P. W. (1976) A non-hydrostatic mesoscale model. *Quarterly Journal Royal Meteorological Society* **102**, 277–296.

## References

TENNEKES, H. (1988) Numerical weather prediction: Illusions of security, tales of imperfection. *Weather* **43,** 165–170.

THORNES, J. E. (1985) Thermal mapping, road danger warnings and ice on roads. *Proceedings of the Second International Road Weather Conference,* Danish Ministry of Transport, Copenhagen, 19 pp.

THORNES, J. E. (1987) *Impact of Climate and Weather on Transport.* An unpublished paper presented in typescript to the workshop on *The Implications of Climatic Variability for Industry in the United Kingdom,* University of Birmingham, 16 pp.

WAISTER, P. D. (1971) Wind shelters improve soft fruit yields. *The Grower* (June 5th) pp 1358–1360.

WILSON, J. W. (1970) Integration of radar and rainfall data for improved rainfall measurements. *Journal of Applied Meteorology* **8,** 489–497.

WORLD METEOROLOGICAL ORGANISATION (1983) *Guide to Meteorological Instruments and Methods of Observation.* 5th Edition. Commission on Instruments and Methods of Observation, Report No. 8, WMO, Geneva.

WORRALL, T. P. & PITTS, J. B. (1987) *Winter Chaos—Can We Buy Our Way Out of It (Railways)?* An Unpublished paper presented in the Introductory Notes to the seminar on *Winter Chaos,* Institution of Civil Engineers, London, pp 17–21.

ZHANG, O-L. & FRITSCH, J. M. (1986) A case study of the sensitivity of numerical simulation of mesoscale convective systems to varying initial conditions. *Monthly Weather Review* **114,** 2418–2431.